Model-Based Clustering, Classification, and Density Estimation Using mclust in R

"The book gives an excellent introduction to using the R package mclust for mixture modeling with (multivariate) Gaussian distributions as well as covering the supervised and semi-supervised aspects. A thorough introduction to the theoretic concepts is given, the software implementation described in detail, and the application shown on many examples. I particularly enjoyed the in-depth discussion of different visualization methods."

- Bettina Grün, WU (Vienna University of Economics and Business), Austria

Model-based clustering and classification methods provide a systematic statistical approach to clustering, classification, and density estimation via mixture modeling. The model-based framework allows the problems of choosing or developing an appropriate clustering or classification method to be understood within the context of statistical modeling. The **mclust** package for the statistical environment R is a widely adopted platform implementing these model-based strategies. The package includes both summary and visual functionality, complementing procedures for estimating and choosing models.

Key features of the book:

- An introduction to the model-based approach and the **mclust** R package
- A detailed description of **mclust** and the underlying modeling strategies
- An extensive set of examples, color plots, and figures along with the R code for reproducing them
- Supported by a companion website including the R code to reproduce the examples and figures presented in the book, errata, and other supplementary material

Model-Based Clustering, Classification, and Density Estimation Using mclust in R is accessible to quantitatively trained students and researchers with a basic understanding of statistical methods, including inference and computing. In addition to serving as a reference manual for **mclust**, the book will be particularly useful to those wishing to employ these model-based techniques in research or applications in statistics, data science, clinical research, social science, and many other disciplines.

Chapman & Hall/CRC
The R Series

Series Editors

John M. Chambers, Department of Statistics, Stanford University, California, USA
Torsten Hothorn, Division of Biostatistics, University of Zurich, Switzerland
Duncan Temple Lang, Department of Statistics, University of California, Davis, USA
Hadley Wickham, RStudio, Boston, Massachusetts, USA

Recently Published Titles

R for Conservation and Development Projects: A Primer for Practitioners
Nathan Whitmore

Using R for Bayesian Spatial and Spatio-Temporal Health Modeling
Andrew B. Lawson

Engineering Production-Grade Shiny Apps
Colin Fay, Sébastien Rochette, Vincent Guyader, and Cervan Girard

Javascript for R
John Coene

Advanced R Solutions
Malte Grosser, Henning Bumann, and Hadley Wickham

Event History Analysis with R, Second Edition
Göran Broström

Behavior Analysis with Machine Learning Using R
Enrique Garcia Ceja

Rasch Measurement Theory Analysis in R: Illustrations and Practical Guidance for Researchers and Practitioners
Stefanie Wind and Cheng Hua

Spatial Sampling with R
Dick R. Brus

Crime by the Numbers: A Criminologist's Guide to R
Jacob Kaplan

Analyzing US Census Data: Methods, Maps, and Models in R
Kyle Walker

ANOVA and Mixed Models: A Short Introduction Using R
Lukas Meier

Tidy Finance with R
Stefan Voigt, Patrick Weiss, and Christoph Scheuch

Deep Learning and Scientific Computing with R torch
Sigrid Keydana

Model-Based Clustering, Classification, and Density Estimation Using mclust in R
Luca Scrucca, Chris Fraley, T. Brendan Murphy, and Adrian E. Raftery

For more information about this series, please visit: https://www.crcpress.com/Chapman--HallCRC-The-R-Series/book-series/CRCTHERSER

Model-Based Clustering, Classification, and Density Estimation Using mclust in R

Luca Scrucca, Chris Fraley, T. Brendan Murphy, and Adrian E. Raftery

CRC Press
Taylor & Francis Group
Boca Raton London New York

CRC Press is an imprint of the
Taylor & Francis Group, an **informa** business

A CHAPMAN & HALL BOOK

First edition published 2023
by CRC Press
6000 Broken Sound Parkway NW, Suite 300, Boca Raton, FL 33487-2742

and by CRC Press
4 Park Square, Milton Park, Abingdon, Oxon, OX14 4RN

CRC Press is an imprint of Taylor & Francis Group, LLC

ISBN: 978-1-032-23496-0 (hbk)
ISBN: 978-1-032-23495-3 (pbk)
ISBN: 978-1-00-327796-5 (ebk)

DOI: 10.1201/9781003277965

Typeset in LM Roman
by KnowledgeWorks Global Ltd.

Publisher's note: This book has been prepared from camera-ready copy provided by the authors.

Contents

List of Figures

List of Tables

List of Examples

Preface

Model-based clustering and classification methods provide a systematic statistical modeling framework for cluster analysis and classification. The model-based approach has gained in popularity because it allows the problems of choosing or developing an appropriate clustering or classification method to be understood within the context of statistical modeling.

mclust is a widely-used software package for the statistical environment R (https://www.r-project.org). It provides functionality for model-based clustering, classification, and density estimation, including methods for summarizing and visualizing the estimated models.

This book aims at giving a detailed overview of **mclust** and its features. A description of the modeling underpinning the software is provided, along with examples of its usage. In addition to serving as a reference manual for **mclust**, the book will be particularly useful to readers who plan to employ these model-based techniques in their research or applications.

Who is this book for?

The book is written to appeal to quantitatively trained readers from a wide range of backgrounds. An understanding of basic statistical methods, including statistical inference and statistical computing, is required. Throughout the book, examples and code are used extensively in an expository style to demonstrate the use of **mclust** for model-based clustering, classification, and density estimation.

Additionally, the book can serve as a reference for courses in multivariate analysis, statistical learning, machine learning, and data mining. It would also be a useful reference for advanced quantitative courses in application areas, including social sciences, physical sciences, and business.

Companion website

A companion website for this book is available at

`https://mclust-org.github.io/book`

The website contains the R code to reproduce the examples and figures presented in the book, errata and various supplementary material.

Software information and conventions

The R session information when compiling this book is shown below:

```
sessionInfo()
## R version 4.2.2 (2022-10-31)
## Platform: x86_64-apple-darwin17.0 (64-bit)
## Running under: macOS Big Sur ... 10.16
##
## Matrix products: default
##
## locale:
## [1] en_US.UTF-8/en_US.UTF-8/en_US.UTF-8/C/en_US.UTF-8/en_US.UTF-8
##
## attached base packages:
## [1] stats     graphics  utils     datasets  grDevices methods
## [7] base
##
## other attached packages:
## [1] knitr_1.42    mclust_6.0.0
##
## loaded via a namespace (and not attached):
## [1] compiler_4.2.2 cli_3.6.0     tools_4.2.2    highr_0.10
## [5] xfun_0.37      rlang_1.0.6   evaluate_0.20
```

Every R input command starts on a new line without any additional prompt (as > or +). The corresponding output is shown on lines starting with two hashes ##, as it can be seen from the R session information above. Package names are in bold text (e.g., **mclust**), and inline code and filenames are formatted in a typewriter font (e.g., `data("iris", package = "datasets")`). Function names are followed by parentheses (e.g., `Mclust()`).

About the authors

Luca Scrucca ⓘ https://orcid.org/0000-0003-3826-0484
Associate Professor of Statistics at Università degli Studi di Perugia, his research interests include: mixture models, model-based clustering and classification, statistical learning, dimension reduction methods, genetic and evolutionary algorithms. He is currently Associate Editor for the *Journal of Statistical Software* and *Statistics and Computing*. He has developed and he is the maintainer of several high profile R packages available on The Comprehensive R Archive Network (CRAN). His webpage is at http://www.stat.unipg.it/luca.

Chris Fraley
Most recently a research staff member at Tableau, she previously held research positions in Statistics at the University of Washington and at Insightful from its early days as Statistical Sciences. She has contributed to computational methods in a number of areas of applied statistics, and is the principal author of several widely-used R packages. She was the originator (at Statistical Sciences) of numerical functions such as nlminb that have long been available in the R core stats package.

T. Brendan Murphy ⓘ https://orcid.org/0000-0002-5668-7046
Professor of Statistics at University College Dublin, his research interests include: model-based clustering, classification, network modeling, and latent variable modeling. He is interested in applications in social science, political science, medicine, food science, and biology. He served as Associate Editor for the journal *Statistics and Computing*, he is currently Editor for the *Annals of Applied Statistics* and Associate Editor for *Statistical Analysis and Data Mining*. His webpage is at http://mathsci.ucd.ie/~brendan

Adrian Raftery ⓘ https://orcid.org/0000-0002-6589-301X
Boeing International Professor of Statistics and Sociology, and Adjunct Professor of Atmospheric Sciences at the University of Washington, Seattle. He is also a faculty affiliate of the Center for Statistics and the Social Sciences and the Center for Studies in Demography and Ecology at University of Washington. He was one of the founding researchers in model-based clustering, having published in the area since 1984. His research interests include: model-based clustering, Bayesian statistics, social network analysis, and statistical demography. He is interested in applications in social, environmental, biological, and health sciences. He is a member of the U.S. National Academy of Sciences and was identified by Thomson-Reuter as the most cited researcher in mathematics in the world for the decade 1995—2005. He served as Editor of the *Journal of the American Statistical Association* (JASA). His webpage is at http://www.stat.washington.edu/raftery.

Acknowledgments

The idea for writing this book arose during one of the yearly meetings of the *Working Group on Model-Based Clustering*, which constitutes a small but very active place for scholars from all over the world interested in mixture modeling. We thank all of the participants for providing the stimulating environment in which we started this project.

We are also fortunate to have benefited from a thorough review contributed by Bettina Grün, a leading expert in mixture modeling.

We have many others to thank for their contributions to **mclust** as users, collaborators, and developers. Thanks also to the R core team, and to those responsible for the many packages we have leveraged.

The development of the **mclust** package was supported over many years by the U.S. Office of Naval Research (ONR), and we acknowledge the encouragement and enthusiasm of our successive ONR program officers, Julia Abrahams and Wendy Martinez.

Chris Fraley is indebted to Tableau for supporting her efforts as co-author.

Brendan Murphy's research was supported by the Science Foundation Ireland (SFI) Insight Research Centre (SFI/12/RC/2289_P2), Vistamilk Research Centre (16/RC/3835) and Collegium de Lyon — Institut d'Études Avancées, Université de Lyon.

Adrian Raftery's research was supported by the Eunice Kennedy Shriver National Institute for Child Health and Human Development (NICHD) under grant number R01 HD070936, by the Blumstein-Jordan and Boeing International Professorships at the University of Washington, and by the Fondation des Sciences Mathématiques de Paris (FSMP) and Université Paris-Cité.

Finally, special thanks to Rob Calver, Senior Publisher at Chapman & Hall/CRC, for his encouragement and enthusiastic support for this book.

October 2022

Luca, Chris, Brendan, and Adrian

Perugia, Seattle, Dublin, and Seattle

1

Introduction

Model-based clustering and classification methods provide a systematic statistical modeling framework for cluster analysis and classification, allowing the problems of choosing or developing an appropriate clustering or classification method to be understood within the context of statistical modeling. This chapter introduces model-based clustering and finite mixture modeling by providing historical background and an overview of the **mclust** software for the R statistical environment.

1.1 Model-Based Clustering and Finite Mixture Modeling

Cluster analysis is the automatic categorization of objects into groups based on their measured characteristics. This book is about model-based approaches to cluster analysis and their implementation. It is also about how these approaches can be applied to classification and density estimation.

The grouping of objects based on what they have in common is a human universal and is inherent in language itself. Plato formalized the idea with his Theory of Forms, and Aristotle may have been the first to implement it empirically, classifying animals into groups based on their characteristics in his *History of Animals*. This was extended by Linneaus in the 18th century with his system of biological classification or taxonomy of animals and plants.

Aristotle and Linneaus classified objects subjectively, but cluster analysis is something more, using systematic numerical methods. It seems to be have been invented by Czekanowski (1909), using measures of similarity between objects based on multiple measurements. In the 1950s, there was renewed interest in the area due to the invention of new hierarchical clustering methods, including the single, average and complete linkage methods.

These methods are heuristic and algorithmic, leaving several key questions unanswered, such as: Which clustering method should we use? How many clusters are there? How should we treat outliers — objects that do not fall into any group? How should we assess uncertainty about an estimated clustering partition?

Early clustering developments were largely separate from mainstream statistics, much of which was based on a probability model for the data. The main statistical model for clustering is a finite mixture model, in which each group is modeled by its own probability distribution. The first method of this kind was latent class analysis for multivariate discrete data, developed by Paul Lazarsfeld (Lazarsfeld, 1950a,b).

The most popular model for clustering continuous-valued data is the mixture of multivariate normal distributions, introduced for this purpose by John Wolfe (Wolfe, 1963, 1965, 1967, 1970), which is the main focus of this book. This modeling approach reduces the questions we mentioned to standard statistical problems such as parameter estimation and model selection. Different clustering methods often correspond approximately to different mixture models, and so choosing a method can often be done by selecting the best model. Each number of clusters corresponds to a different mixture model, so that choosing the number of clusters also becomes a model selection problem. Outliers can also be accounted for in the probability model.

1.2   mclust

mclust (Fraley et al., 2022) is a popular R (R Core Team, 2022) package for model-based clustering, classification, and density estimation based on finite Gaussian mixture models (GMMs). It provides an integrated approach to finite mixture models, with functions that combine model-based hierarchical clustering, the EM (Expectation-Maximization) algorithm for mixture estimation (Dempster et al., 1977; McLachlan and Krishnan, 2008), and several tools for model selection. A variety of covariance structures and cross-component constraints are available (see Section 2.2.1). Also included are functions for performing individual E and M steps, for simulating data from each available model, and for displaying and visualizing fitted models along with the associated clustering, classification, and density estimation results. The most recent versions of the package provide dimension reduction for visualization, resampling-based inference, additional model selection criteria, and more options for initializing the EM algorithm. A web page for the **mclust** package, and other related R packages, can be accessed at the URL https://mclust-org.github.io.

mclust was first developed in 1991 by Chris Fraley and Adrian Raftery for model-based hierarchical clustering with geometric constraints (Banfield and Raftery, 1993), and subsequently expanded to include constrained Gaussian mixture modeling via EM (Celeux and Govaert, 1995). This extended the original methodology in John Wolfe's NORMIX software (Wolfe, 1967) by including a range of more parsimonious and statistically efficient models, and by adding methods for choosing the number of clusters and the best model, and for identifying outliers. **mclust** was originally implemented in the S-Plus statistical computing environment, calling Fortran for numerical operations, and using the BLAS and LAPACK numerical subroutines, which at the time were not widely available outside of Fortran. It was later ported to R by Ron Wehrens.

Earlier versions of the package were described in Fraley and Raftery (1999), Fraley and Raftery (2003), and Fraley et al. (2012). More recent versions of the package are described in Scrucca et al. (2016). The current version at the time of writing is 6.0.0.

mclust offers a comprehensive strategy for clustering, classification, and density estimation, and is in increasingly high demand as shown in Figure 1.1. This graph shows the monthly downloads from the RStudio CRAN mirror over the last few years, with figures calculated using the database provided by the R package **cranlogs** (Csárdi, 2019).

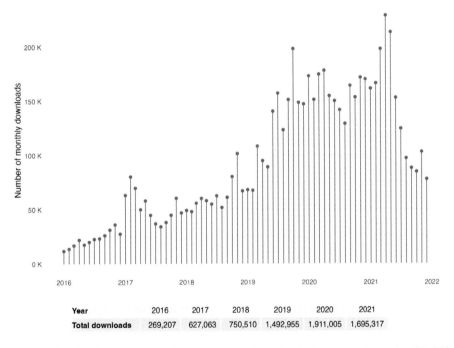

Year	2016	2017	2018	2019	2020	2021
Total downloads	269,207	627,063	750,510	1,492,955	1,911,005	1,695,317

FIGURE 1.1: Number of **mclust** monthly downloads from the RStudio CRAN mirror over the last few years and total downloads by year.

1.3   Overview

mclust currently includes the following features:

- Normal (Gaussian) mixture modeling via EM for fourteen specifications of covariance structures and cross-component constraints (structured GMMs).

- Simulation from all available model specifications.

- Model-based clustering that combines model fitting via structured GMMs with model selection using BIC and other options.

- Density estimation using GMMs.

- Methods for combining mixture components for clustering.

- Discriminant analysis (classification) based on structured GMMs (EDDA, MclustDA) and semi-supervised classification.

- Dimension reduction methods for model-based clustering and classification.

- Displays, including uncertainty plots, random projections, contour and perspective plots, classification plots, and density curves in one and two dimensions.

mclust is a package for the R language available on CRAN at `https://cran.r-project.org/web/packages/mclust` and licensed under the GPL `https://www.gnu.org/licenses/gpl.html`. There are ready to install versions, both in binary and in source format, for several machines and operating systems. The simplest way to install the latest version of **mclust** from CRAN is to use the following command from the R console:

```
install.packages("mclust")
```

Once the package is installed, it can be loaded into an R session using the command:

```
library("mclust")
##                        __          __
##      ____ ___ _____/ /_ _____/ /_
##    / __ `__ \/ __/ / / / / __/ __/
##   / / / / / / /__/ / /_/ (__  ) /_
##  /_/ /_/ /_/\___/_/\__,_/____/\__/   version 6.0.0
## Type 'citation("mclust")' for citing this R package in publications.
```

Color-Blind Accessibility

mclust includes various options to accommodate color-blind users. For details, see Section 6.6.

1.4 Organization of the Book

The book is organized as follows. Chapter 2 gives a general introduction to finite mixture models and the special case of Gaussian mixture models (GMMs) which is emphasized in this book. It describes common methods for parameter estimation and model selection.

Chapter 3 describes the general methodology for model-based clustering, including model estimation and selection. It discusses algorithm initialization at some length, as this is a major issue for model-based clustering.

Chapter 4 describes mixture-based classification or supervised learning. It describes various ways of assessing classifier performance, and also discusses semi-supervised classification, in which only some of the training data have known labels.

Chapter 5 describes methods for model-based univariate and multivariate density estimation. Chapter 6 describes ways of visualizing the results of model-based clustering and discusses the underlying considerations.

Finally, Chapter 7 concludes by discussing a range of other issues, including accounting for outliers and noise. It describes Bayesian methods for avoiding the singularities that can arise in mixture modeling by adding a prior. It also describes two approaches to the common situation where clusters are not Gaussian: using an entropy criterion to combine GMM mixture components and identifying connected components. Simulation from mixture models is also discussed briefly, as well as handling large datasets, high-dimensional data, and missing data.

2

Finite Mixture Models

This chapter gives a general introduction to finite mixture models and the special case of Gaussian mixture models (GMMs) which is emphasized in this book. It describes common methods for parameter estimation and model selection. In particular, the maximum likelihood approach is presented and the EM algorithm for maximum likelihood estimation is detailed. The Gaussian case is discussed at length. We introduce a parsimonious covariance decomposition that allows one to regularize the estimation procedure. The maximum a posteriori procedure is described as a way to obtain probabilistic clustering. Methods for model selection based on information criteria and likelihood ratio testing are presented. Finally, inference on parameters is discussed by adopting a resampling-based approach.

2.1   Finite Mixture Models

Mixture models encompass a powerful set of statistical tools for cluster analysis, classification, and density estimation. They provide a widely-used family of models that have proved to be an effective and computationally convenient way to model data arising in many fields, from agriculture to astronomy, economics to medicine, marketing to bioinformatics, among others. Details of finite mixture models and their applications can be found in Titterington et al. (1985); McLachlan and Basford (1988); McLachlan and Peel (2000); Bishop (2006, Chapter 9); Frühwirth-Schnatter (2006); McNicholas (2016); Bouveyron et al. (2019). In this book our interest in mixture models will be mostly in their use for statistical learning problems, mostly unsupervised, but also supervised.

A mixture distribution is a probability distribution obtained as a convex linear combination[1] of probability density functions[2]. The individual distributions that are combined to form the mixture distribution are called *mixture components*, and the weights associated with each component are called *mixture*

---

[1] Roughly speaking, a convex linear combination is a weighted sum of terms with non-negative weights that sum to one.

[2] A probability density function may be defined with respect to an appropriate measure on $\mathbb{R}^d$, which can be the Lebesgue measure, a counting measure, or a combination of the two, depending on the context.

weights or *mixture proportions*. The number of mixture components is often restricted to being finite, although in some cases it may be countably infinite. The general form of the density of a *finite mixture distribution* for a d-dimensional random vector $\boldsymbol{x}$ can be written in the form

$$\sum_{k=1}^{G} \pi_k f_k(\boldsymbol{x}; \boldsymbol{\theta}_k), \qquad (2.1)$$

where G is the number of mixture components, $f_k(\cdot)$ is the density of the kth component of the mixture (with $k = 1, \ldots, G$), the π_k's are the mixture weights ($\pi_k > 0$ and $\sum_{k=1}^{G} \pi_k = 1$), and $\boldsymbol{\theta}_k$ represents the parameters of the kth density component. Typically, the component densities are taken to be known up to the parameters $\boldsymbol{\theta}_k$ for $k = 1, \ldots, G$.

It is usually assumed that all of the densities $f_k(\boldsymbol{x}; \boldsymbol{\theta}_k)$ belong to the same parametric family of distributions, but with different parameters. However, in some circumstances, different parametric forms are appropriate, such as in zero-inflated models where a component is introduced for modeling an excess of zeros. In Section 7.1, we introduce an additional component with a Poisson distribution to account for noise in the data.

Mixture distributions can be used to model a wide variety of random phenomena, in particular those that cannot be adequately described by a single parametric distribution. For instance, they are suitable for dealing with *unobserved heterogeneity*, which occurs when a sample is drawn from a statistical population without knowledge of the presence of underlying sub-populations. In this case, the mixture components can be seen as the densities of the sub-populations, and the mixing weights are the proportions of each sub-population in the overall population.

EXAMPLE 2.1: Using Gaussian mixtures to explain fish length heterogeneity

Consider the fish length measurements (in inches) for 256 snappers attributed to Cassie (1954). The data, available as Snapper in the R package **FSAdata** (Ogle, 2022), show a certain amount of heterogeneity with the presence of several modes. A possible explanation is that the fish belong to different age groups, but age is hard to measure, so no information is collected about this characteristic. Mixtures of Gaussian distributions with up to four components were fitted to this data, and the resulting mixture densities are shown in Figure 2.1.

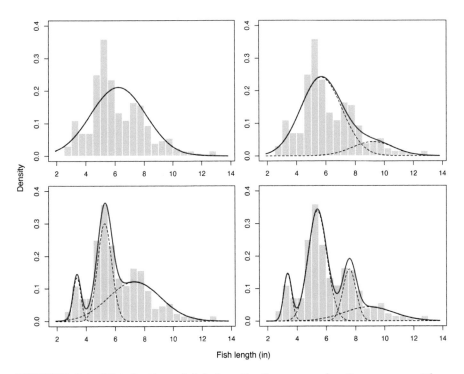

FIGURE 2.1: Distribution of fish lengths for a sample of snappers with estimated Gaussian mixtures from 1 up to 4 number of mixture components. Dashed lines represent weighed component densities, solid lines the mixture densities.

2.1.1   Maximum Likelihood Estimation and the EM Algorithm

Given a random sample of observations $x_1, x_2, \ldots, x_n$, the likelihood of a finite mixture model with G components is given by

$$L(\boldsymbol{\Psi}) = \prod_{i=1}^{n} \left\{ \sum_{k=1}^{G} \pi_k f_k(\boldsymbol{x}_i; \boldsymbol{\theta}_k) \right\},$$

where $\boldsymbol{\Psi} = (\pi_1, \ldots, \pi_{G-1}, \boldsymbol{\theta}_1, \ldots, \boldsymbol{\theta}_G)$ are the parameters to be estimated. The corresponding log-likelihood is

$$\ell(\boldsymbol{\Psi}) = \sum_{i=1}^{n} \log \left\{ \sum_{k=1}^{G} \pi_k f_k(\boldsymbol{x}_i; \boldsymbol{\theta}_k) \right\}. \tag{2.2}$$

The maximum likelihood estimate (MLE) of $\boldsymbol{\Psi}$ is defined as a stationary point of the likelihood in the interior of the parameter space, and is thus a root of the likelihood equation $\partial \ell(\boldsymbol{\Psi})/\partial \boldsymbol{\Psi} = \boldsymbol{0}$ corresponding to a finite local maximum. However, the log-likelihood in (2.2) is hard to maximize directly, even numerically (see McLachlan and Peel, 2000, Section 2.8.1). As a consequence, mixture models are usually fitted by reformulating the mixture problem as an incomplete-data problem within the EM framework.

The *Expectation-Maximisation (EM) algorithm* (Dempster et al., 1977) is a general approach to maximum likelihood estimation when the data can be seen as the realization of multivariate observations $(\boldsymbol{x}_i, \boldsymbol{z}_i)$ for $i = 1, \ldots, n$, where the $\boldsymbol{x}_i$ are observed and the $\boldsymbol{z}_i$ are latent, unobserved variables. In the case of finite mixture models, $\boldsymbol{z}_i = (z_{i1}, \ldots, z_{iG})^\top$, where

$$
z_{ik} = \begin{cases} 1 & \text{if } \boldsymbol{x}_i \text{ belongs to the } k\text{th component of the mixture,} \\ 0 & \text{otherwise.} \end{cases}
$$

Under the i.i.d. (*independent and identically distributed*) assumption for the random variables $(\boldsymbol{x}_i, \boldsymbol{z}_i)$, the *complete-data likelihood* is given by

$$
L_C(\boldsymbol{\Psi}) = \prod_{i=1}^{n} f(\boldsymbol{x}_i, \boldsymbol{z}_i; \boldsymbol{\Psi}) = \prod_{i=1}^{n} p(\boldsymbol{z}_i) f(\boldsymbol{x}_i; \boldsymbol{z}_i, \boldsymbol{\Psi}).
$$

Assuming that the $\boldsymbol{z}_i$ are i.i.d. according to the multinomial distribution with probabilities $(\pi_1, \ldots, \pi_G)$, it follows that

$$
p(\boldsymbol{z}_i) \propto \prod_{k=1}^{G} \pi_k^{z_{ik}},
$$

and

$$
f(\boldsymbol{x}_i; \boldsymbol{z}_i, \boldsymbol{\Psi}) = \prod_{k=1}^{G} f_k(\boldsymbol{x}_i; \boldsymbol{\theta}_k)^{z_{ik}}.
$$

Thus the complete-data log-likelihood is given by

$$
\ell_C(\boldsymbol{\Psi}) = \sum_{i=1}^{n} \sum_{k=1}^{G} z_{ik} \{\log \pi_k + \log f_k(\boldsymbol{x}_i; \boldsymbol{\theta}_k)\},
$$

where $\boldsymbol{\Psi} = (\pi_1, \ldots, \pi_{G-1}, \boldsymbol{\theta}_1, \ldots, \boldsymbol{\theta}_G)$ are the unknown parameters.

The *EM algorithm* is an iterative procedure whose objective function at each iteration is the conditional expectation of the complete-data log-likelihood, the Q-function, which for finite mixtures takes the form:

$$
Q(\boldsymbol{\Psi}; \boldsymbol{\Psi}^{(t)}) = \sum_{i=1}^{n} \sum_{k=1}^{G} \widehat{z}_{ik}^{(t)} \{\log \pi_k + \log f_k(\boldsymbol{x}_i; \boldsymbol{\theta}_k)\},
$$

where $\widehat{z}_{ik}^{(t)} = \mathrm{E}(z_{ik} = 1 | \boldsymbol{x}_i, \boldsymbol{\Psi}^{(t)})$, the estimated conditional probability that $\boldsymbol{x}_i$ belongs to the kth component at iteration t of the EM algorithm.

In general, the EM algorithm for finite mixtures consists of the following steps:

- Initialization: set $t = 0$ and choose initial values for the parameters, $\boldsymbol{\Psi}^{(0)}$.

- E-step — estimate the latent component memberships:

$$\widehat{z}_{ik}^{(t)} = \widehat{\Pr}(z_{ik} = 1 | \boldsymbol{x}_i, \widehat{\boldsymbol{\Psi}}^{(t)}) = \frac{\pi_k^{(t)} f_k(\boldsymbol{x}_i; \boldsymbol{\theta}_k^{(t)})}{\sum_{j=1}^{G} \pi_j^{(t)} f_j(\boldsymbol{x}_i; \boldsymbol{\theta}_j^{(t)})}.$$

- M-step — obtain the updated parameter estimates:

$$\boldsymbol{\Psi}^{(t+1)} = \arg \max_{\boldsymbol{\Psi}} Q(\boldsymbol{\Psi}; \boldsymbol{\Psi}^{(t)}).$$

Note that, for finite mixture models,

$$\pi_k^{(t+1)} = \frac{\sum_{i=1}^{n} \widehat{z}_{ik}^{(t)}}{n}.$$

- If convergence criteria are not satisfied, set $t = t + 1$ and perform another E-step followed by an M-step.

As an alternative to specifying initial values for the parameters, the EM algorithm for finite mixture models can be invoked with an initial assignment of observations to mixture components. The latter is equivalent to starting EM from the M-step with $\widehat{z}_{ik}^{(0)}$ set to 1 if the ith observation is assigned to component k, and 0 otherwise. More details on initialization are given in Section 2.2.3.

Properties of the EM algorithm have been extensively studied in the literature; for a review see McLachlan and Krishnan (2008). Some of the main advantages are the following:

- Unless a stationary point of the log-likelihood has been reached, each EM iteration increases the log-likelihood. Although the likelihood surface for a GMM is unbounded wherever a covariance is singular, EM tends to converge to finite local maxima.

- In many cases of practical interest, the E-steps and M-steps are more tractable in terms of implementation than direct maximization of the log-likelihood, and the cost per iteration is often relatively low.

- For mixture models, probabilities are guaranteed to remain in $[0, 1]$, and it is possible to implement EM for Gaussian mixture models (GMMs) in such a way that the covariance matrices cannot have negative eigenvalues.

Unfortunately, there are also drawbacks such as the following:

- The resulting parameter estimates can be highly dependent on their initial values, as well as on the convergence criteria.

- Convergence may be difficult to assess: not only can the asymptotic rate of convergence be slow, but progress can also be slow even when the current value is far away from a stationary point.

- The advantages of EM may not be fully realized due to numerical issues in the implementation.

- An estimate of the covariance matrix of the parameter estimates (needed to assess uncertainty) is not available as a byproduct of the EM computations. Methods have been developed to overcome this, such as the resampling approach described in Section 2.4.

2.1.2   Issues in Maximum Likelihood Estimation

When computing the MLE of a finite mixture model, some potential problems may arise. The first issue is that the mixture likelihood may be unbounded (see McLachlan and Peel, 2000, Sections 2.2 and 2.5). For example, a global maximum does not exist for Gaussian mixture models (GMMs) with unequal covariance matrices (McLachlan and Peel, 2000, Section 3.8.1). Optimization methods may diverge and fail to converge to a finite local optimum. Imposing cross-cluster constraints, as discussed for GMMs in Section 2.2.1, reduces the chances of encountering unboundedness during optimization. Another approach, which can be combined with constraints, is to add a prior distribution for regularization (see Section 7.2). Further alternatives are discussed by Hathaway (1985); Ingrassia and Rocci (2007); García-Escudero et al. (2015).

Another issue is that the likelihood surface often has many local maxima. If an iterative optimization method does converge to a local maximum, the corresponding parameter values will depend on how that method was initialized. Initialization strategies for the EM algorithm for GMMs are discussed in Section 2.2.3. Moreover, as mentioned above, not only can the asymptotic rate of convergence be slow, but progress can also be slow away from the optimum. As result, convergence criteria, which are typically confined to absolute or relative changes in the log-likelihood and/or parameters, may be satisfied at a non-stationary point.

Identifiability of the mixture components poses another potential problem. The log-likelihood in (2.2) is maximized for any permutation of the order of the components (the *label switching problem*). This is not usually a problem with the EM algorithm for finite mixture models, but it can be a serious problem for Bayesian approaches that rely on sampling from the posterior distribution. For further details and remedies, see Frühwirth-Schnatter (2006).

2.2 Gaussian Mixture Models

Mixtures of Gaussian distributions are the most popular model for continuous data, that is, numerical data that can theoretically be measured in infinitely small units. Gaussian mixture models (GMMs) are widely used in statistical learning, pattern recognition, and data mining (Celeux and Govaert, 1995; Fraley and Raftery, 2002; Stahl and Sallis, 2012).

The probability density function of a GMM can be written as

$$f(\boldsymbol{x}; \boldsymbol{\Psi}) = \sum_{k=1}^{G} \pi_k \phi(\boldsymbol{x}; \boldsymbol{\mu}_k, \boldsymbol{\Sigma}_k), \tag{2.3}$$

where $\phi(\cdot)$ is the multivariate Gaussian density function with mean $\boldsymbol{\mu}_k$ and covariance matrix $\boldsymbol{\Sigma}_k$:

$$\phi(\boldsymbol{x}; \boldsymbol{\mu}_k, \boldsymbol{\Sigma}_k) = \frac{1}{\sqrt{(2\pi)^d |\boldsymbol{\Sigma}_k|}} \exp\left\{-\frac{1}{2}(\boldsymbol{x} - \boldsymbol{\mu})^\top \boldsymbol{\Sigma}_k^{-1}(\boldsymbol{x} - \boldsymbol{\mu})\right\}.$$

In this case the vector of unknown parameters is given by $\boldsymbol{\Psi} = (\pi_1, \ldots, \pi_{G-1}, \boldsymbol{\mu}_1, \ldots, \boldsymbol{\mu}_G, \text{vech}\{\boldsymbol{\Sigma}_1\}, \ldots, \text{vech}\{\boldsymbol{\Sigma}_G\})^\top$, where $\text{vech}\{\cdot\}$ is an operator that forms a vector by extracting the unique elements of a symmetric matrix. Alternatively, the covariance matrix can be parameterized by its Cholesky factor. This latter parameterization is used for most of the models in **mclust**.

The GMM is a flexible model that can serve different purposes. In this book we will mainly discuss applications of Gaussian mixtures in clustering (Chapter 3), classification (Chapter 4), and density estimation (Chapter 5).

2.2.1 Parsimonious Covariance Decomposition

Data generated by a GMM are characterized by groups or clusters centered at the mean $\boldsymbol{\mu}_k$, with higher density for points closer to the mean. Isosurfaces of constant density are ellipsoids whose geometric characteristics (such as volume, shape, and orientation) are determined by the covariance matrices $\boldsymbol{\Sigma}_k$. The number of parameters per mixture component grows quadratically with the dimensionality of the data for the GMM with unrestricted component covariance matrices. Introducing cross-component constraints may help to avoid issues with near-singular covariance estimates (see Section 2.1.2).

Geometric characteristics of the GMM components can be controlled by imposing constraints on the covariance matrices through the eigen-decomposition (Banfield and Raftery, 1993; Celeux and Govaert, 1995):

$$\boldsymbol{\Sigma}_k = \lambda_k \boldsymbol{U}_k \boldsymbol{\Delta}_k \boldsymbol{U}_k^\top, \tag{2.4}$$

where $\lambda_k = |\boldsymbol{\Sigma}_k|^{1/d}$ is a scalar controlling the *volume*, $\boldsymbol{\Delta}_k$ is a diagonal matrix controlling the *shape*, such that $|\mathbf{\Delta}_k| = 1$ and with the normalized eigenvalues of $\mathbf{\Sigma}_k$ in decreasing order, and $\mathbf{U}_k$ is an orthogonal matrix of eigenvectors of $\mathbf{\Sigma}_k$ controlling the *orientation*.

Characteristics of component distributions, such as volume, shape, and orientation, are usually estimated from the data, and can be allowed to vary between clusters, or constrained to be the same for all clusters (Murtagh and Raftery, 1984; Flury, 1988; Banfield and Raftery, 1993; Celeux and Govaert, 1995). Accordingly, $(\lambda_k, \mathbf{\Delta}_k, \mathbf{U}_k)$ can be treated as independent sets of parameters. Components that share the same value of λ will have the same volume, while those that share the same value of $\mathbf{\Delta}$ will have the same shape, and those that have the same value of $\mathbf{U}$ will have the same orientation.

TABLE 2.1: Parameterizations of the covariance matrix $\mathbf{\Sigma}_k$ for multidimensional data.

Label	Model	Distribution	Volume	Shape	Orientation
EII	$\lambda \mathbf{I}$	Spherical	Equal	Equal	—
VII	$\lambda_k \mathbf{I}$	Spherical	Variable	Equal	—
EEI	$\lambda \mathbf{\Delta}$	Diagonal	Equal	Equal	Coordinate axes
VEI	$\lambda_k \mathbf{\Delta}$	Diagonal	Variable	Equal	Coordinate axes
EVI	$\lambda \mathbf{\Delta}_k$	Diagonal	Equal	Variable	Coordinate axes
VVI	$\lambda_k \mathbf{\Delta}_k$	Diagonal	Variable	Variable	Coordinate axes
EEE	$\lambda \mathbf{U} \mathbf{\Delta} \mathbf{U}^\top$	Ellipsoidal	Equal	Equal	Equal
VEE	$\lambda_k \mathbf{U} \mathbf{\Delta} \mathbf{U}^\top$	Ellipsoidal	Variable	Equal	Equal
EVE	$\lambda \mathbf{U} \mathbf{\Delta}_k \mathbf{U}^\top$	Ellipsoidal	Equal	Variable	Equal
VVE	$\lambda_k \mathbf{U} \mathbf{\Delta}_k \mathbf{U}^\top$	Ellipsoidal	Variable	Variable	Equal
EEV	$\lambda \mathbf{U}_k \mathbf{\Delta} \mathbf{U}_k^\top$	Ellipsoidal	Equal	Equal	Variable
VEV	$\lambda_k \mathbf{U}_k \mathbf{\Delta} \mathbf{U}_k^\top$	Ellipsoidal	Variable	Equal	Variable
EVV	$\lambda \mathbf{U}_k \mathbf{\Delta}_k \mathbf{U}_k^\top$	Ellipsoidal	Equal	Variable	Variable
VVV	$\lambda_k \mathbf{U}_k \mathbf{\Delta}_k \mathbf{U}_k^\top$	Ellipsoidal	Variable	Variable	Variable

Table 2.1 lists the 14 possible models that can be obtained for multidimensional data by varying these geometric characteristics of the component distributions. The reference label and its component-covariance model and distributional form, followed by the corresponding characteristics of volume, shape, and orientation, are given for each model. In Figure 2.2 these geometric characteristics are represented graphically for a bivariate case with three groups.

In the nomenclature adopted in this book and in the **mclust** software, E and V indicate, respectively, equal and variable characteristics across groups, while I is the identity matrix. For example, EVI denotes a model in which the volumes of all clusters are equal (E), the shapes of the clusters may vary (V), and the orientation is the identity (I). According to this model specification, clusters have diagonal covariances with orientation parallel to the coordinate axes. In the one-dimensional case there are just two possible models: E for equal variance, and V for varying variance. In all cases, the parameters associated with characteristics designated by E or V are to be determined from the data, as discussed in the next section.

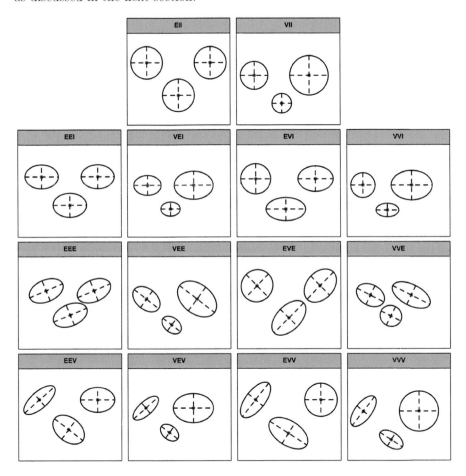

FIGURE 2.2: Ellipses of isodensity for each of the 14 Gaussian models parameterized by the eigen-decomposition of the component covariances for the case of three groups in two dimensions. The first row shows the two spherical models ⋆II, followed by the four diagonal models ⋆⋆I, then the four equal-orientation models ⋆⋆E, and the four varying-orientation models ⋆⋆V.

2.2.2   EM Algorithm for Gaussian Mixtures

For Gaussian component densities, $\phi(\boldsymbol{x}_i; \boldsymbol{\mu}_k, \boldsymbol{\Sigma}_k)$, the log-likelihood can be written as

$$\ell(\boldsymbol{\Psi}) = \sum_{i=1}^{n} \log \left\{ \sum_{k=1}^{G} \pi_k \phi(x_i; \boldsymbol{\mu}_k, \boldsymbol{\Sigma}_k) \right\},$$

where $\boldsymbol{\Psi}$ is the set of parameters to be estimated as described above. The complete-data log-likelihood is then given by

$$\ell_C(\boldsymbol{\Psi}) = \sum_{i=1}^{n} \sum_{k=1}^{G} z_{ik} \left\{ \log \pi_k + \log \phi(\boldsymbol{x}_i; \boldsymbol{\mu}_k, \boldsymbol{\Sigma}_k) \right\}.$$

The EM algorithm for GMMs follows the general approach outlined in Section 2.1.1, with the following steps (omitting the dependence on iteration t for clarity of exposition):

- E-step:

$$\widehat{z}_{ik} = \frac{\widehat{\pi}_k \phi(\boldsymbol{x}_i; \widehat{\boldsymbol{\mu}}_k, \widehat{\boldsymbol{\Sigma}}_k)}{\sum_{g=1}^{G} \widehat{\pi}_g \phi(\boldsymbol{x}_i; \widehat{\boldsymbol{\mu}}_g, \widehat{\boldsymbol{\Sigma}}_g)},$$

- M-step:

$$\widehat{\pi}_k = \frac{n_k}{n} \quad \text{and} \quad \widehat{\boldsymbol{\mu}}_k = \frac{\sum_{i=1}^{n} \widehat{z}_{ik} \boldsymbol{x}_i}{n_k}, \quad \text{where } n_k = \sum_{i=1}^{n} \widehat{z}_{ik}.$$

Estimation of $\boldsymbol{\Sigma}_k$ depends on the adopted parameterization for the component-covariance matrices. Some simple cases are listed in the table below, where $\boldsymbol{W}_k = \sum_{i=1}^{n} \widehat{z}_{ik}(\boldsymbol{x}_i - \widehat{\boldsymbol{\mu}}_k)(\boldsymbol{x}_i - \widehat{\boldsymbol{\mu}}_k)^{\top}$, and $\boldsymbol{W} = \sum_{k=1}^{G} \boldsymbol{W}_k$.

	$\boldsymbol{\Sigma}_k$ model	$\widehat{\boldsymbol{\Sigma}}_k$
EII	λI	$\dfrac{\mathrm{tr}(\boldsymbol{W})}{dn} \boldsymbol{I}$
VII	$\lambda_k I$	$\dfrac{\mathrm{tr}(\boldsymbol{W}_k)}{dn_k} \boldsymbol{I}$
EEE	$\lambda \boldsymbol{U} \boldsymbol{\Delta} \boldsymbol{U}^{\top}$	$\dfrac{\boldsymbol{W}}{n}$
VVV	$\lambda_k \boldsymbol{U}_k \boldsymbol{\Delta}_k \boldsymbol{U}_k^{\top}$	$\dfrac{\boldsymbol{W}_k}{n_k}$

Celeux and Govaert (1995) discuss the M-step for all 14 models and provide iterative methods for the 5 models (VEI, VEE, VEV, EVE, VVE) for which the

M-step does not have a closed form. An alternative based on MM (Minorize-Maximization) optimization is used in **mclust** for the M-step in the EVE and VVE models (Browne and McNicholas, 2014).

Table 2.2 gives the complexity, measured by the number of parameters to be estimated, and indicates whether the M-step is in closed form (CF), or requires an iterative procedure (IP). Figure 2.3 shows the increasing complexity of GMMs as a function of the number of mixture components and number of variables for the available models. Clearly, the number of parameters to be estimated grows much faster for more flexible models.

TABLE 2.2: Number of estimated parameters and M-step for GMMs with different covariance parameterizations for multidimensional data.

Label	Model	Number of parameters	M-step
EII	$\lambda \boldsymbol{I}$	$(G-1) + Gd + 1$	CF
VII	$\lambda_k \boldsymbol{I}$	$(G-1) + Gd + G$	CF
EEI	$\lambda \boldsymbol{\Delta}$	$(G-1) + Gd + d$	CF
VEI	$\lambda_k \boldsymbol{\Delta}$	$(G-1) + Gd + G + (d-1)$	IP
EVI	$\lambda \boldsymbol{\Delta}_k$	$(G-1) + Gd + 1 + G(d-1)$	CF
VVI	$\lambda_k \boldsymbol{\Delta}_k$	$(G-1) + Gd + G + G(d-1)$	CF
EEE	$\lambda \boldsymbol{U\Delta U}^{\top}$	$(G-1) + Gd + 1 + (d-1) + d(d-1)/2$	CF
VEE	$\lambda_k \boldsymbol{U\Delta U}^{\top}$	$(G-1) + Gd + G + (d-1) + d(d-1)/2$	IP
EVE	$\lambda \boldsymbol{U\Delta}_k \boldsymbol{U}^{\top}$	$(G-1) + Gd + 1 + G(d-1) + d(d-1)/2$	IP
VVE	$\lambda_k \boldsymbol{U\Delta}_k \boldsymbol{U}^{\top}$	$(G-1) + Gd + G + G(d-1) + d(d-1)/2$	IP
EEV	$\lambda \boldsymbol{U}_k \boldsymbol{\Delta U}_k^{\top}$	$(G-1) + Gd + 1 + (d-1) + Gd(d-1)/2$	CF
VEV	$\lambda_k \boldsymbol{U}_k \boldsymbol{\Delta U}_k^{\top}$	$(G-1) + Gd + G + (d-1) + Gd(d-1)/2$	IP
EVV	$\lambda \boldsymbol{U}_k \boldsymbol{\Delta}_k \boldsymbol{U}_k^{\top}$	$(G-1) + Gd + 1 + G(d-1) + Gd(d-1)/2$	CF
VVV	$\lambda_k \boldsymbol{U}_k \boldsymbol{\Delta}_k \boldsymbol{U}_k^{\top}$	$(G-1) + Gd + G + G(d-1) + Gd(d-1)/2$	CF

The number of parameters to be estimated includes $(G-1)$ for the mixture weights and Gd for the component means for all models. The number of covariance parameters varies with the model. In the M-step column, CF indicates that the M-step is available in closed form, while IP indicates that the M-step requires an iterative procedure.

Figure 2.4 shows some steps of the EM algorithm used for fitting a two-component unrestricted Gaussian mixture model to the Old Faithful data (Azzalini and Bowman, 1990). More details about the dataset are given in Section 3.3.1. Here the EM algorithm is initialized by a random partition. Points are marked according to the maximum a posteriori (MAP) classification (2.2.4) that assigns each $\boldsymbol{x}_i$ to the mixture component with the largest posterior conditional probability. Ellipses show the distribution of the current Gaussian components. Initially the component densities overlap to a large extent, but after only a few iterations of the EM algorithm the separation between the components clearly emerges. This is also reflected in the separation of the observed data points into two clusters.

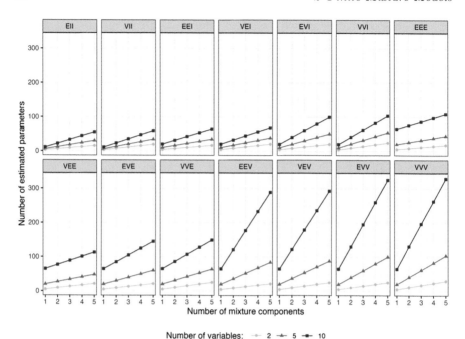

FIGURE 2.3: Number of GMM estimated parameters as a function of the number of mixture components, for different numbers of variables and cross-component covariance constraints.

2.2.3 Initialization of EM Algorithm

The EM algorithm is an iterative, strictly hill-climbing procedure whose performance can depend strongly on the starting point because the finite mixture likelihood surface tends to have multiple modes. Thus, initialization of the EM algorithm is often crucial, although no method suggested in the literature uniformly outperforms the others. Nevertheless, the EM algorithm is usually able to produce sensible results when started from reasonable starting values (Wu, 1983; Everitt et al., 2001, p. 150).

In the case of Gaussian mixtures, several approaches, both stochastic and deterministic, are available for selecting an initial partition of the observations or an initial estimate of the parameters. Broadly speaking, there are two general approaches for starting the EM algorithm.

In the first approach, the EM algorithm is initialized using a set of randomly selected parameters. For instance, a simple strategy is based on generating several candidates by drawing parameter values uniformly at random over the feasible parameter region. Alternatively, membership probabilities can be drawn at random over the unit simplex of dimension equal to the number of mixture components. Since the random-starts strategy has a fair chance of

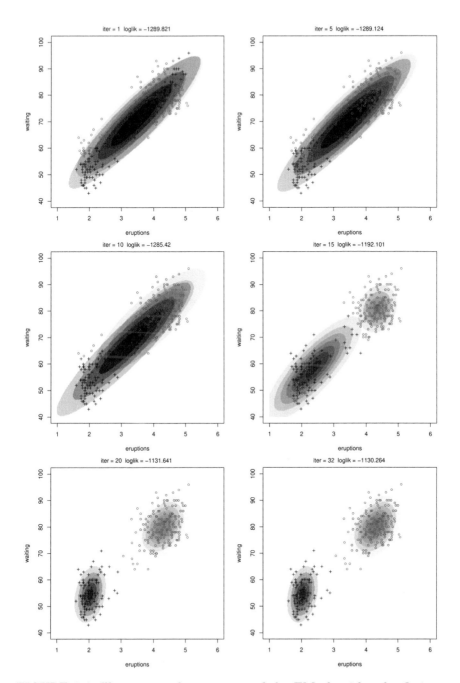

FIGURE 2.4: Illustration of some steps of the EM algorithm for fitting a two-component Gaussian mixture model to the Old Faithful data.

failing to provide good initial starting values, a common suggestion is to run the EM algorithm with several random starts and choose the one resulting in the highest log-likelihood.

Another stochastic initialization scheme is the *emEM* strategy proposed by Biernacki et al. (2003). This uses several short runs of the EM algorithm initialized with valid random starts as parameter estimates until an overall number of total iterations is exhausted. Then, the one with the highest log-likelihood is chosen to be the initializer for a long-running EM, which runs until the usual strict convergence criteria are met. The R package **Rmixmod** (Langrognet et al., 2022) uses this by default. However, emEM is computationally intensive and suffers from the same issues mentioned above for random starts, although to a lesser extent.

Another approach to initializing the EM algorithm is based on the partition obtained from some other clustering algorithm, such as k-means or hierarchical clustering. In this case, the final classification is used to start the EM algorithm from the M-step. However, there are drawbacks associated with the use of these partitioning algorithms for initializing EM. For example, some have their own initialization issues, and some have a tendency to artificially impose specific shapes or patterns on clusters.

In the **mclust** R package, the EM algorithm is initialized using the partitions obtained from model-based agglomerative hierarchical clustering (MBAHC). In this approach, k clusters are obtained from a large number of smaller clusters by recursively merging the two clusters that yield the maximum likelihood of a probability model over all possible merges. Banfield and Raftery (1993) proposed using the Gaussian classification likelihood as the underlying criterion. For the simplest model with equal, spherical covariance matrices, this is the same criterion that underlies the classical sum-of-squares method. Fraley (1998) showed how the structure of some Gaussian models can be exploited to yield efficient regularized algorithms for agglomerative hierarchical clustering. Further details are given in Sections 3.6 and 3.7.

2.2.4   Maximum A Posteriori (MAP) Classification

Given a dataset $\mathcal{X} = \{x_1, x_2, \ldots, x_n\}$, a hard partition of the observed data points into G clusters, denoted as $\mathcal{C} = \{C_1, C_2, \ldots, C_G\}$ such that $C_k \cap C_g = \emptyset$ (for $k \neq g$) and $\bigcup_{k=1}^{G} C_k = \mathcal{X}$, is straightforward to obtain in finite mixture modeling.

Once a GMM has been successfully fitted and the MLEs of the parameters obtained, a maximum a posteriori (MAP) procedure can be applied, assigning each x_i to the mixture component with the largest posterior conditional probability:

$$x_i \in C_{k^*} \qquad \text{with} \quad k^* = \arg\max_k \widehat{z}_{ik},$$

where

$$\widehat{z}_{ik} = \frac{\widehat{\pi}_k \phi(\boldsymbol{x}_i; \widehat{\boldsymbol{\mu}}_k, \widehat{\boldsymbol{\Sigma}}_k)}{\displaystyle\sum_{g=1}^{G} \widehat{\pi}_g \phi(\boldsymbol{x}; \widehat{\boldsymbol{\mu}}_g, \widehat{\boldsymbol{\Sigma}}_g)} \tag{2.5}$$

is the posterior conditional probability of an observation i coming from mixture component k ($k = 1, \ldots, G$). A measure of classification uncertainty for each data point can also be computed as

$$u_i = 1 - \max_k \widehat{z}_{ik},$$

which falls within the interval $[0, 1]$. Values close to zero indicate a low level of uncertainty in the classification of the corresponding observation.

2.3 Model Selection

A central question in finite mixture modeling is that of determining how many components should be included in the mixture. In GMMs we need also to decide which covariance parameterization to adopt. Both questions can be addressed by model selection criteria, such as the Bayesian information criterion (BIC) or the integrated complete-data likelihood (ICL) criterion. The selection of the number of mixture components or clusters can also be done by formal hypothesis testing.

2.3.1 Information Criteria

Information criteria are usually based on penalized forms of the likelihood. In general, as the log-likelihood increases with the addition of more parameters in a statistical model, a penalty term for the number of estimated parameters is included to account for the model complexity (Claeskens and Hjort, 2008; Konishi and Kitagawa, 2008).

Let $\boldsymbol{x}_1, \ldots, \boldsymbol{x}_n$ be a random sample of n independent observations. Consider a parametric family of density functions $\{f(\boldsymbol{x}; \boldsymbol{\theta}); \boldsymbol{\theta} \in \boldsymbol{\Theta}\}$, for which the log-likelihood can be computed as $\ell(\boldsymbol{\theta}; \boldsymbol{x}) = \log L(\boldsymbol{\theta}; \boldsymbol{x}) = \sum_{i=1}^{n} \log f(\boldsymbol{x}_i; \boldsymbol{\theta})$, where $\widehat{\boldsymbol{\theta}}$ is the MLE (the value that maximizes the log-likelihood). The *Bayesian information criterion* (BIC) , originally introduced by Schwartz (1978), is a popular criterion for model selection that penalizes the log-likelihood by introducing a penalty term:

$$\text{BIC} = 2\ell(\widehat{\boldsymbol{\theta}}; \boldsymbol{x}) - \nu_{\boldsymbol{\theta}} \log(n),$$

where $\ell(\widehat{\boldsymbol{\theta}}; \boldsymbol{x})$ is the maximized log-likelihood, n is the sample size, and $\nu_{\boldsymbol{\theta}}$ is the number of parameters to be estimated.

Kass and Raftery (1995) showed that, assuming prior unit information, BIC provides an approximation to the Bayes factor for comparing two competing models, say $\mathcal{M}_1$ and $\mathcal{M}_2$:

$$2 \log B_{12} \approx \text{BIC}_{\mathcal{M}_1} - \text{BIC}_{\mathcal{M}_2} = \Delta_{12}.$$

Assuming that $\mathcal{M}_2$ has the smaller BIC value, the strength of the evidence against it can be summarized as follows:

Δ_{12}	Evidence to favor $\mathcal{M}_1$ over $\mathcal{M}_2$
$0 - 2$	Not worth more than a bare mention
$2 - 6$	Positive
$6 - 10$	Strong
> 10	Very Strong

For a review of BIC, its derivation, its properties and applications see Neath and Cavanaugh (2012).

The BIC is a widely adopted criterion for model selection in finite mixture models, both for density estimation (Roeder and Wasserman, 1997) and for clustering (Fraley and Raftery, 1998). For mixture models, it takes the following form:

$$\text{BIC}_{\mathcal{M},G} = 2\ell_{\mathcal{M},G}(\widehat{\boldsymbol{\Psi}}; \boldsymbol{x}) - \nu_{\mathcal{M},G} \log(n),$$

where $\ell_{\mathcal{M},G}(\widehat{\boldsymbol{\Psi}}; \boldsymbol{x})$ is the log-likelihood at the MLE $\widehat{\boldsymbol{\Psi}}$ for model $\mathcal{M}$ with G components, n is the sample size, and $\nu_{\mathcal{M},G}$ is the number of parameters to be estimated. The model $\mathcal{M}$ and number of components G are chosen so as to maximize $\text{BIC}_{\mathcal{M},G}$. [3] Keribin (2000) showed that BIC is consistent for choosing the number of components in a mixture model, under the assumption that the likelihood is bounded. Although GMM likelihoods have an infinite spike wherever one or more covariances is singular, BIC is nevertheless often used for model selection among GMMs.

The BIC tends to select the number of mixture components needed to approximate the density, rather than the number of clusters as such. For this reason, other criteria have been proposed for model selection in clustering, like the *integrated complete-data likelihood* (ICL) criterion (Biernacki et al., 2000):

$$\text{ICL}_{\mathcal{M},G} = \text{BIC}_{\mathcal{M},G} + 2 \sum_{i=1}^{n} \sum_{k=1}^{G} c_{ik} \log(\widehat{z}_{ik}),$$

where $\widehat{z}_{ik}$ is the conditional probability that $\boldsymbol{x}_i$ arises from the kth mixture component from equation (2.5), and $c_{ik} = 1$ if the ith observation is assigned to cluster k, i.e. $\boldsymbol{x}_i \in C_k$, and 0 otherwise. ICL penalizes the BIC through an *entropy* term which measures the overlap between clusters. Provided that the clusters do not overlap too much, ICL has shown good performance in selecting the number of clusters, with a preference for solutions with well-separated groups.

---

[3] BIC is often taken to have the opposite sign in the literature, and is thus minimized for model selection in those cases.

2.3.2 Likelihood Ratio Testing

In addition to the information criteria just mentioned, the choice of the number of components in a mixture model for a specific component-covariance parameterization can be carried out by likelihood ratio testing (LRT); see McLachlan and Rathnayake (2014) for a review.

Suppose we want to test the null hypothesis $G = G_0$ against the alternative $G = G_1$ for some $G_1 > G_0$, so that

$$H_0 : G = G_0$$
$$H_1 : G = G_1.$$

Usually, $G_1 = G_0 + 1$, so a common procedure is to keep adding components sequentially. Let $\widehat{\boldsymbol{\Psi}}_{G_j}$ be the MLE of $\boldsymbol{\Psi}$ calculated under $H_j : G = G_j$ (for $j = 0, 1$). The likelihood ratio test statistic (LRTS) can be written as

$$\text{LRTS} = -2 \log\{L(\widehat{\boldsymbol{\Psi}}_{G_0})/L(\widehat{\boldsymbol{\Psi}}_{G_1})\} = 2\{\ell(\widehat{\boldsymbol{\Psi}}_{G_1}) - \ell(\widehat{\boldsymbol{\Psi}}_{G_0})\},$$

where large values of LRTS provide evidence against the null hypothesis. For mixture models, however, standard regularity conditions do not hold for the null distribution of the LRTS to have its usual chi-squared distribution (McLachlan and Peel, 2000, Chapter 6). As a result, the significance of the LRT is often assessed using a resampling approach in order to obtain a p-value. McLachlan (1987) proposed using the bootstrap to obtain the null distribution of the LRTS. The bootstrap procedure is the following:

1. a bootstrap sample $\boldsymbol{x}_b^*$ is generated by simulating from the fitted model under the null hypothesis with G_0 components, namely, from the GMM distribution with the vector of unknown parameters replaced by MLEs obtained from the original data under H_0;

2. the test statistic LRTS_b^* is computed for the bootstrap sample $\boldsymbol{x}_b^*$ after fitting GMMs with G_0 and G_1 number of components;

3. steps 1. and 2. are replicated B times, say $B = 999$, to obtain the bootstrap null distribution of LRTS^*.

A bootstrap-based approximation of the p-value may then be computed as

$$p\text{-value} \approx \frac{1 + \sum_{b=1}^{B} I(\text{LRTS}_b^* \geq \text{LRTS}_{\text{obs}})}{B + 1},$$

where LRTS_{obs} is the test statistic computed on the observed sample, and $I(\cdot)$ denotes the indicator function, which is equal to 1 if its argument is true and 0 otherwise.

2.4   Resampling-Based Inference

The EM algorithm for Gaussian mixtures provides an efficient way to obtain parameter estimates (see Section 2.2.2). However, as already mentioned in Section 2.1.1, the EM algorithm does not provide estimates of the uncertainty associated with the parameter estimates. Likelihood-based inference in mixture models is usually addressed through either information-based methods or resampling (McLachlan and Peel, 2000; McLachlan and Krishnan, 2008).

In information-based methods (Meng and Rubin, 1991), the covariance matrix of the MLE $\widehat{\boldsymbol{\Psi}}$ is approximated by the inverse of the observed information matrix $I^{-1}(\widehat{\boldsymbol{\Psi}})$:

$$\mathrm{Cov}(\widehat{\boldsymbol{\Psi}}) \approx I^{-1}(\widehat{\boldsymbol{\Psi}}).$$

Although valid asymptotically (Boldea and Magnus, 2009), "the sample size n has to be very large before the asymptotic theory applies to mixture models" (McLachlan and Peel, 2000, p. 42). Indeed, Basford et al. (1997) found that standard errors obtained using the expected or the observed information matrix are unstable unless the sample size is very large. For these reasons, they advocate the use of a resampling approach based on the bootstrap.

The *bootstrap* (Efron, 1979) is a general, widely applicable, powerful technique for obtaining an approximation to the sampling distribution of a statistic of interest. The bootstrap distribution is approximated by drawing a large number of samples (*bootstrap samples*) from the empirical distribution, by resampling with replacement from the observed data (*nonparametric bootstrap*), or from a parametric distribution with unknown parameters substituted by the corresponding estimates (*parametric bootstrap*).

Let $\widehat{\boldsymbol{\Psi}}$ be the estimate of a set of GMM parameters $\boldsymbol{\Psi}$ for a given model $\mathcal{M}$ and number of mixture components G. A bootstrap estimate of the corresponding standard errors can be obtained as follows:

- Obtain the bootstrap distribution for the parameters of interest by:

  1. drawing a sample of size n with replacement from the empirical distribution $(\boldsymbol{x}_1, \dots, \boldsymbol{x}_n)$ to form the bootstrap sample $(\boldsymbol{x}_1^*, \dots, \boldsymbol{x}_n^*)$;
  2. fitting a GMM $(\mathcal{M}, G)$ to get the bootstrap estimates $\widehat{\boldsymbol{\Psi}}^*$;
  3. replicating steps 1–2 a large number of times, say B, to obtain $\widehat{\boldsymbol{\Psi}}_1^*, \widehat{\boldsymbol{\Psi}}_2^*, \dots, \widehat{\boldsymbol{\Psi}}_B^*$ estimates from B resamples.

- An approximate covariance matrix for the parameter estimates is then

$$\mathrm{Cov}_{\mathrm{boot}}(\widehat{\boldsymbol{\Psi}}) = \frac{1}{B-1} \sum_{b=1}^{B} (\widehat{\boldsymbol{\Psi}}_b^* - \overline{\widehat{\boldsymbol{\Psi}}}^*)(\widehat{\boldsymbol{\Psi}}_b^* - \overline{\widehat{\boldsymbol{\Psi}}}^*)^\top$$

where $\overline{\widehat{\boldsymbol{\Psi}}}^* = \dfrac{1}{B} \displaystyle\sum_{b=1}^{B} \widehat{\boldsymbol{\Psi}}_b^*$.

- The bootstrap standard errors for the parameter estimates $\widehat{\boldsymbol{\Psi}}$ are computed as the square root of the diagonal elements of the bootstrap covariance matrix:

$$\mathrm{se}_{\mathrm{boot}}(\widehat{\boldsymbol{\Psi}}) = \sqrt{\mathrm{diag}(\mathrm{Cov}_{\mathrm{boot}}(\widehat{\boldsymbol{\Psi}}))}.$$

The bootstrap procedure outlined above can also be used for estimating confidence intervals. For instance, bootstrap percentile confidence intervals for any GMM parameter ψ of $\boldsymbol{\Psi}$ are computed as $[\psi^*_{\alpha/2}, \psi^*_{1-\alpha/2}]$, where ψ^*_q is the qth quantile (or the $100q$th percentile) of the bootstrap distribution $(\widehat{\psi}^*_1, \ldots, \widehat{\psi}^*_B)$.

Different resampling methods have been considered for Gaussian mixtures, such as the parametric bootstrap (Basford et al., 1997), the nonparametric bootstrap (McLachlan and Peel, 2000), and the jackknife (Efron, 1979, 1982). A generalization of the nonparametric bootstrap is the *weighted likelihood bootstrap* (Newton and Raftery, 1994), which assigns random (positive) weights to sample observations. The weights are obtained from a uniform Dirichlet distribution, by sampling from n independent standard exponential distributions and then rescaling by their average. The weighted likelihood bootstrap can also be viewed as a generalized Bayesian bootstrap. The weighted likelihood bootstrap may yield benefits when one or more components have small mixture proportions. In that case, a nonparametric bootstrap sample may have no representatives of them, whereas the weighted likelihood bootstrap always has representatives of all groups.

For a recent review and comparison of these resampling approaches to inference in finite mixture models see O'Hagan et al. (2019).

3

Model-Based Clustering

This chapter describes the general methodology for Gaussian model-based clustering in **mclust**, including model estimation and selection. Several data examples are presented, in both the multivariate and univariate case. EM algorithm initialization is discussed at some length, as this is a major issue for model-based clustering. Model-based hierarchical agglomerative clustering is also presented, and the corresponding implementation in **mclust** is shown. The chapter concludes by providing some details on performing single E- and M- steps, and on control parameters used by the EM functions in **mclust**.

3.1 Gaussian Mixture Models for Cluster Analysis

Cluster analysis refers to a broad set of multivariate statistical methods and techniques which seek to identify homogeneous subgroups of cases in a dataset, that is, to partition similar observations into meaningful and useful *clusters*. Cluster analysis is an instance of *unsupervised learning*, since the presence and number of clusters may not be known a priori, nor is case labeling available.

Many approaches to clustering have been proposed in the literature for exploring the underlying group structure of data. Traditional methods are combinatorial in nature, and either *hierarchical* (agglomerative or divisive) or *non-hierarchical* (for example, k-means). Although some are related to formal statistical models, they typically are based on heuristic procedures which make no explicit assumptions about the structure of the clusters. The choice of clustering method, similarity measures, and interpretation have tended to be informal and often subjective.

Model-based clustering (MBC) is a probabilistic approach to clustering in which each cluster corresponds to a mixture component described by a probability distribution with unknown parameters. The type of distribution is often specified a priori (most commonly Gaussian), whereas the model structure (including the number of components) remains to be determined by parameter estimation and model-selection techniques. Parameters can be estimated by maximum likelihood, using, for instance, the EM algorithm, or by Bayesian approaches. These model-based methods are increasingly favored over heuristic clustering methods. Moreover, the one-to-one relationship between mixture components and clusters can be relaxed, as discussed in Section 7.3.

Model-based clustering can be viewed as a *generative* approach because it attempts to learn generative models from the data, with each model representing one particular cluster. It is possible to simulate from an estimated model, as discussed in Section 7.4, which can be useful for resampling inference, sensitivity, and predictive analysis.

The finite mixture of Gaussian distributions described in Section 2.2 is one of the most flexible classes of models for continuous variables and, as such, it has become a method of choice in a wide variety of settings. Component covariance matrices can be parameterized by eigen-decomposition to impose geometric constraints, as described in Section 2.2.1. Constraining some but not all of the quantities in (2.4) to be equal between clusters yields parsimonious and easily interpretable models, appropriate for diverse clustering situations. Once the parameters of a Gaussian mixture model (GMM) are estimated via EM, a partition $\mathcal{C} = \{C_1, \ldots, C_G\}$ of the data into G clusters can be derived by assigning each observation $\boldsymbol{x}_i$ to the component with the largest conditional probability that $\boldsymbol{x}_i$ arises from that component distribution — the *MAP principle* discussed in Section 2.2.4.

The parameterization includes, but is not restricted to, well-known covariance models that are associated with various criteria for hierarchical clustering: equal-volume spherical variance (sum of squares criterion) (Ward, 1963), constant variance (Friedman and Rubin, 1967), and unconstrained variance (Scott and Symons, 1971). Moreover, k-means, although it does not make an explicit assumption about the structure of the data, is related to the GMM with spherical, equal volume components (Celeux and Govaert, 1995).

3.2 Clustering in mclust

The main function to fit GMMs for clustering is called `Mclust()`. This function requires a numeric matrix or data frame, with n observations along the rows and d variables along the columns, to be supplied as the `data` argument. In the univariate case ($d = 1$), the data can be input as a vector. The number of mixture components G, and the model name corresponding to the eigen-decomposition discussed in Section 2.2.1, can also be specified. By default, all of the available models are estimated for up to $G = 9$ mixture components, but this number can be increased by specifying the argument `G` accordingly.

Table 3.1 lists the models currently implemented in **mclust**, both in the univariate and multivariate cases. Options for specification of a noise component and a prior distribution are also shown; these will be discussed in Sections 7.1 and 7.2, respectively.

TABLE 3.1: Models available in the **mclust** package for hierarchical clustering (HC), Gaussian mixture models (GMM) fitted by the EM algorithm, with noise component or prior (✓ indicates availability).

Label	Model	Distribution	HC	GMM	GMM + noise	GMM + prior
Single component						
X	σ^2	Univariate		✓	✓	✓
XII	$\mathrm{diag}(\sigma^2,\ldots,\sigma^2)$	Spherical		✓	✓	✓
XXI	$\mathrm{diag}(\sigma_1^2,\ldots,\sigma_d^2)$	Diagonal		✓	✓	✓
XXX	$\boldsymbol{\Sigma}$	Ellipsoidal		✓	✓	✓
Multi-components						
E	σ^2	Univariate	✓	✓	✓	✓
V	σ_k^2	Univariate	✓	✓	✓	✓
EII	$\lambda\boldsymbol{I}$	Spherical	✓	✓	✓	✓
VII	$\lambda_k\boldsymbol{I}$	Spherical	✓	✓	✓	✓
EEI	$\lambda\boldsymbol{\Delta}$	Diagonal		✓	✓	✓
VEI	$\lambda_k\boldsymbol{\Delta}$	Diagonal		✓	✓	✓
EVI	$\lambda\boldsymbol{\Delta}_k$	Diagonal		✓	✓	✓
VVI	$\lambda_k\boldsymbol{\Delta}_k$	Diagonal		✓	✓	✓
EEE	$\lambda\boldsymbol{U}\boldsymbol{\Delta}\boldsymbol{U}^\top$	Ellipsoidal	✓	✓	✓	✓
VEE	$\lambda_k\boldsymbol{U}\boldsymbol{\Delta}\boldsymbol{U}^\top$	Ellipsoidal		✓	✓	
EVE	$\lambda\boldsymbol{U}\boldsymbol{\Delta}_k\boldsymbol{U}^\top$	Ellipsoidal		✓	✓	
VVE	$\lambda_k\boldsymbol{U}\boldsymbol{\Delta}_k\boldsymbol{U}^\top$	Ellipsoidal		✓	✓	
EEV	$\lambda\boldsymbol{U}_k\boldsymbol{\Delta}\boldsymbol{U}_k^\top$	Ellipsoidal		✓	✓	✓
VEV	$\lambda_k\boldsymbol{U}_k\boldsymbol{\Delta}\boldsymbol{U}_k^\top$	Ellipsoidal		✓	✓	✓
EVV	$\lambda\boldsymbol{U}_k\boldsymbol{\Delta}_k\boldsymbol{U}_k^\top$	Ellipsoidal		✓	✓	
VVV	$\lambda_k\boldsymbol{U}_k\boldsymbol{\Delta}_k\boldsymbol{U}_k^\top$	Ellipsoidal	✓	✓	✓	✓

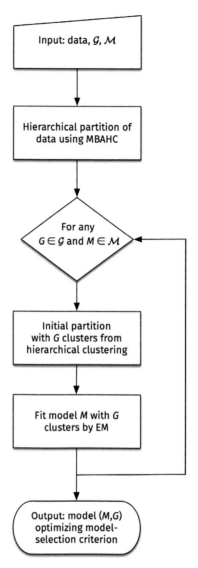

FIGURE 3.1: Flow chart of the model-based clustering approach implemented in **mclust**. $\mathcal{G}$ indicates the set containing the number of mixture components, by default $\mathcal{G} = \{1, 2, \ldots, 9\}$. $\mathcal{M}$ is the set of within-cluster covariance models, by default all of the available models listed in Table 3.1. BIC is the default of several model-selection options.

The diagram in Figure 3.1 presents the flow chart of the model-based clustering approach implemented in the function Mclust() of the **mclust** package.

Given numerical data, a model-based agglomerative hierarchical clustering

(MBAHC) procedure is first applied, for which details are given in Section 3.6. For a set $\mathcal{G}$ giving the numbers of mixture components to consider, and a list $\mathcal{M}$ of models with within-cluster covariance eigen-decomposition chosen from Table 3.1, the hierarchical partitions obtained for any number of components in the set $\mathcal{G}$ are used to start the EM algorithm. GMMs are then fit for any combination of elements $G \in \mathcal{G}$ and $M \in \mathcal{M}$. At the end of the procedure, the model with the best value of the model-selection criterion is returned (BIC is the default for model selection; see Section 3.3 for available options).

EXAMPLE 3.1: Clustering diabetes data

The `diabetes` dataset from Reaven and Miller (1979), available in the R package **rrcov** (Todorov, 2022), contains measurements on 145 non-obese adult subjects. A description of the variables is given in the table below.

Variable	Description
rw	Relative weight, expressed as the ratio of actual weight to expected weight, given the subject's height.
fpg	Fasting plasma glucose level.
glucose	Area under the plasma glucose curve.
insulin	Area under the plasma insulin curve.
sspg	Steady state plasma glucose level, a measure of insulin resistance.

The `glucose` and `insulin` measurements were recorded for a three hour oral glucose tolerance test (OGTT). The subjects are clinically classified into three groups: "normal", "overt" diabetes (the most advanced stage), and "chemical" diabetes (a latent stage preceding overt diabetes).

```
data("diabetes", package = "rrcov")
X <- diabetes[, 1:5]
Class <- diabetes$group
table(Class)
## Class
##   normal chemical    overt
##       76       36       33
```

The data can be shown graphically as in Figure 3.2 with the following code:

```
clp <- clPairs(X, Class, lower.panel = NULL)
clPairsLegend(0.1, 0.3, class = clp$class, col = clp$col, pch = clp$pch)
```

The `clPairs()` function is an enhanced version of the base `pairs()` function, which allows data points to be represented by different colors and symbols. The function invisibly returns a list information useful for adding a legend via the `clPairsLegend()` function.

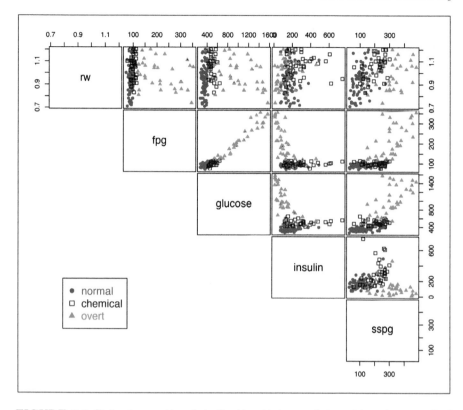

FIGURE 3.2: Pairwise scatterplots for the diabetes data with points marked according to the true classification.

The following command performs a cluster analysis of the diabetes dataset for an unconstrained covariance model (VVV in the nomenclature of Table 2.1 and Table 3.1) with three mixture components:

```
mod <- Mclust(X, G = 3, modelNames = "VVV")
```

Next, a summary of the modeling results is printed:

```
summary(mod)
## ----------------------------------------------------
## Gaussian finite mixture model fitted by EM algorithm
## ----------------------------------------------------
##
## Mclust VVV (ellipsoidal, varying volume, shape, and orientation) model
## with 3 components:
##
##  log-likelihood   n df    BIC     ICL
```

```
##              -2936.8 145 62 -6182.2 -6189.5
##
## Clustering table:
##   1  2  3
## 78 40 27
```

By adding the optional argument parameters = TRUE, a more detailed summary can be obtained, including the estimated parameters:

```
summary(mod, parameters = TRUE)
## ----------------------------------------------------------
## Gaussian finite mixture model fitted by EM algorithm
## ----------------------------------------------------------
##
## Mclust VVV (ellipsoidal, varying volume, shape, and orientation) model
## with 3 components:
##
##  log-likelihood   n df      BIC      ICL
##         -2936.8 145 62  -6182.2  -6189.5
##
## Clustering table:
##   1  2  3
## 78 40 27
##
## Mixing probabilities:
##       1       2       3
## 0.53635 0.27605 0.18760
##
## Means:
##               [,1]      [,2]       [,3]
## rw         0.93884    1.0478    0.98353
## fpg       91.76513  102.9543  236.39448
## glucose  357.21471  508.7962 1127.76651
## insulin  166.14835  298.1248   78.38848
## sspg     107.47991  228.7259  338.06062
##
## Variances:
## [,,1]
##                   rw      fpg   glucose     insulin      sspg
## rw         0.0160109   0.3703    1.7519  -0.0072541    2.5199
## fpg        0.3703007  63.8795  128.5895  29.8548772   58.8162
## glucose    1.7518634 128.5895 2090.3853 269.2839704  408.9926
## insulin   -0.0072541  29.8549  269.2840 2656.1047748  856.5919
## sspg       2.5198642  58.8162  408.9926 856.5919127 2131.0809
## [,,2]
```

```
##                  rw        fpg    glucose    insulin      sspg
## rw          0.011785   -0.25403    -2.4794     1.7088     2.2958
## fpg        -0.254030  210.19848  1035.1415  -242.8201   -74.2975
## glucose    -2.479438 1035.14145  8151.8109 -1301.0449 -1021.8653
## insulin     1.708775 -242.82013 -1301.0449 24803.2354  2202.7878
## sspg        2.295790  -74.29753 -1021.8653  2202.7878  2453.5726
## [,,3]
##                  rw        fpg    glucose    insulin      sspg
## rw          0.013712    -3.1314    -16.683     2.7459     3.2804
## fpg        -3.131374  4910.7530 17669.044  -2208.7550  2630.6654
## glucose   -16.682509 17669.0438 71474.571  -9214.1986  9005.4355
## insulin     2.745916 -2208.7550 -9214.199   2124.2757  -296.8184
## sspg        3.280440  2630.6654  9005.435   -296.8184  6392.4694
```

The following table shows the relationship between the true classes and those defined by the fitted GMM:

```
table(Class, Cluster = mod$classification)
##             Cluster
## Class        1  2  3
##    normal   70  6  0
##    chemical  8 28  0
##    overt     0  6 27
adjustedRandIndex(Class, mod$classification)
## [1] 0.65015
```

Roughly, the first cluster appears to be associated with normal subjects, the second cluster with chemical diabetes, and the last cluster with subjects suffering from overt diabetes. Also shown is the *adjusted Rand index* (ARI; Hubert and Arabie, 1985), which can be used for evaluating a clustering solution. The ARI is a measure of agreement between two partitions, one that is estimated by a statistical procedure ignoring the labeling of the groups and the other that is the true classification. It has zero expected value in the case of a random partition, corresponding to the hypothesis of independent clusterings with fixed marginals, and is bounded above by 1, with higher values representing better partition accuracy. In this case, a better classification could be obtained by including the full range of available covariance models (the default) in the call to Mclust().

The plot() method associated with Mclust() provides for a variety of displays. For instance, the following code produces a plot showing the clustering partition derived from the fitted GMM:

```
plot(mod, what = "classification")
```

The resulting plot is shown in Figure 3.3, with points marked on the basis of the

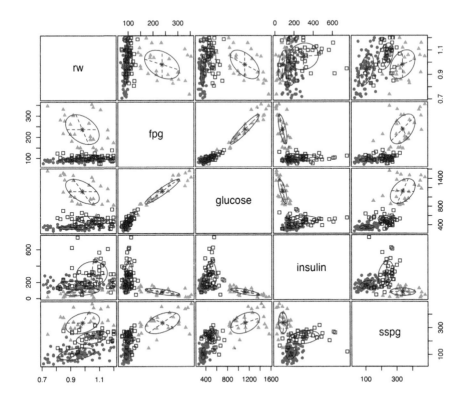

FIGURE 3.3: Scatterplot matrix of variables in the diabetes data with points colored according to the **mclust** classification, and ellipses corresponding to projections of the **mclust** cluster covariances.

MAP classification (see Section 2.2.4), and ellipses showing the projection of the two-dimensional analog of the standard deviation for the cluster covariances of the estimated GMM. The ellipses can be dropped by specifying the optional argument addEllipses = FALSE, and filled ellipses can be obtained with the code:

```
plot(mod, what = "classification", fillEllipses = TRUE)
```

The previous figures showed scatterplot matrices of all pairs of variables. A subset of pairs plots can be specified using the optional argument dimen. For instance, the following code selects the first two variables (see Figure 3.4):

```
plot(mod, what = "classification", dimens = c(3, 4), fillEllipses = TRUE)
```

Uncertainty associated with the MAP classification (see Section 2.2.4) can also be represented graphically as follows:

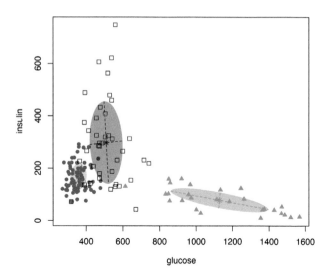

FIGURE 3.4: Scatterplot of a pair of variables in the `diabetes` data with points marked according to the **mclust** classification, and filled ellipses corresponding to **mclust** cluster covariances.

```
plot(mod, dimens = c(3, 4), what = "uncertainty")
```

The resulting plot is shown in Figure 3.5, where the size of the data points reflects the classification uncertainty. Subsets of variables can also be specified for this type of graph.

Plots can be selected interactively by leaving the `what` argument unspecified, in which case the `plot()` function will supply a menu of options. Chapter 6 provides a more detailed description of the graphical capabilities available in **mclust** for further fine-tuning.

EXAMPLE 3.2: Clustering thyroid disease data

Consider the thyroid disease data (Coomans and Broeckaert, 1986) that provides the results of five laboratory tests administered to a sample of 215 patients. The tests are used to predict whether a patient's thyroid can be classified as euthyroidism (normal thyroid gland function), hypothyroidism (underactive thyroid not producing enough thyroid hormone) or hyperthyroidism (overactive thyroid producing and secreting excessive amounts of the free thyroid hormones T3 and/or thyroxine T4). Diagnosis of thyroid function was based on a complete medical record, including anamnesis, scan, and other measurements.

The data is one of several datasets included in the 'Thyroid Disease Data Set' of the UCI Machine Learning Repository (Dua and Graff (2017) — see `https://archive.ics.uci.edu/ml/datasets/thyroid+disease`). The data used

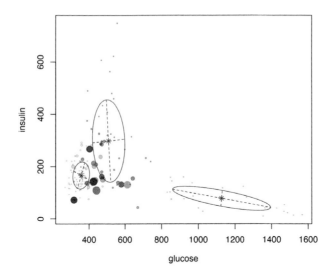

FIGURE 3.5: Scatterplot of a pair of variables in the diabetes data with points marked according to the clustering, and point size reflecting the corresponding uncertainty of the MAP classification for the **mclust** model.

here, which corresponds to new-thyroid.data and new-thyroid.names in the data folder at the UCI site, is available as a dataset in **mclust**. A description of the variables is given in the table below, and a plot of the data is shown in Figure 3.6.

Variable	Description
Diagnosis	Diagnosis of thyroid operation: Hypo, Normal, and Hyper.
RT3U	T3-resin uptake test (percentage).
T4	Total serum thyroxin as measured by the isotopic displacement method.
T3	Total serum triiodothyronine as measured by radioimmuno assay.
TSH	Basal thyroid-stimulating hormone (TSH) as measured by radioimmuno assay.
DTSH	Maximal absolute difference of TSH value after injection of 200 micro grams of thyrotropin-releasing hormone as compared to the basal value.

```
data("thyroid", package = "mclust")
X <- data.matrix(thyroid[, 2:6])
Class <- thyroid$Diagnosis
```

```
clp <- clPairs(X, Class, lower.panel = NULL,
               symbols = c(0, 1, 2),
               colors = c("gray50", "black", "red3"))
clPairsLegend(0.1, 0.3, title = "Thyroid diagnosis:", class = clp$class,
              col = clp$col, pch = clp$pch)
```

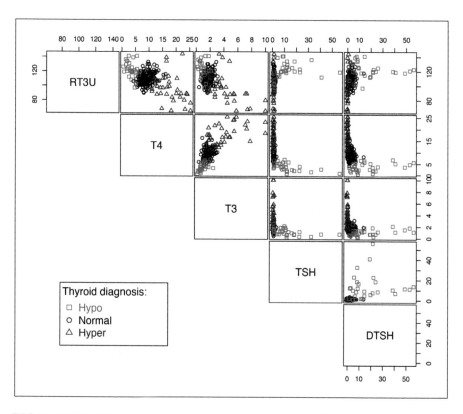

FIGURE 3.6: Pairwise scatterplots showing the classification for the thyroid gland data.

In Example 3.1 we fitted a GMM by specifying both the numbers of mixture components to be considered, using the argument G, and the models for the component-covariance matrices to be used, using the argument modelNames. Users can provide multiple values for both arguments, a vector of integers for G, a vector of character strings for modelNames. By default, G = 1:9, so GMMs with up to nine components are fitted, and all of the available models shown in Table 3.1 are used for modelNames.

Clustering of the thyroid data is obtained using the following R command:

```
mod <- Mclust(X)
```

Since we have provided only the data in the function call, selection of both the number of mixture components and the covariance parameterization is done automatically using the default model selection criterion, BIC. We can inspect the table of BIC values for all the $9 \times 14 = 126$ estimated models as follows:

```
mod$BIC
## Bayesian Information Criterion (BIC):
##       EII      VII      EEI      VEI      EVI      VVI      EEE      VEE
## 1 -7477.4 -7477.4 -6699.7 -6699.7 -6699.7 -6699.7 -6388.4 -6388.4
## 2 -7265.4 -6780.9 -6399.7 -5475.4 -5446.6 -5276.3 -6231.3 -5429.6
## 3 -6971.3 -6404.0 -6081.2 -5471.7 -5110.3 -4777.9 -6228.4 -5273.3
## 4 -6726.5 -6263.1 -5972.8 -5281.3 -5146.0 -4796.8 -6260.7 -5381.4
## 5 -6604.0 -6077.5 -5920.8 -5248.8 -4897.4 -4810.9 -6292.6 -5276.0
## 6 -6574.1 -6002.9 -5893.5 -5230.1 -4945.0 -4854.6 -5946.0 -5291.6
## 7 -6572.8 -5967.0 -6116.2 -5250.0 -4960.5 -4860.0 -5977.3 -5279.3
## 8 -6500.2 -5947.6 -6065.9 -5253.0 -5003.5 -4881.4 -5924.7 -5327.2
## 9 -6370.8 -5904.0 -5579.3 -5244.5 -5089.2 -4876.4 -5696.2 -5273.0
##       EVE      VVE      EEV      VEV      EVV      VVV
## 1 -6388.4 -6388.4 -6388.4 -6388.4 -6388.4 -6388.4
## 2 -5286.3 -5163.6 -5354.9 -5166.4 -5241.5 -5150.8
## 3 -5036.4 -5453.9 -5272.8 -5181.0 -5048.4 -4809.8
## 4 -4974.8 -5478.8 -5129.3 -4902.6 -5061.9 -5175.3
## 5 -5022.6 -5474.2 -5233.9 -4972.1      NA -4959.9
## 6 -4932.2 -4878.3 -5290.6 -5005.4 -5141.0 -5012.5
## 7 -5039.8 -4888.7 -5359.3 -5209.3 -5250.2 -5096.8
## 8 -5086.9 -4909.7 -5376.2 -5070.9 -5347.1 -5129.7
## 9      NA      NA -5287.7 -5091.2      NA      NA
##
## Top 3 models based on the BIC criterion:
##   VVI,3   VVI,4   VVV,3
## -4777.9 -4796.8 -4809.8
```

The missing values (denoted by NA) that appear in the BIC table indicate that a particular model could not be estimated as initialized. This usually happens when a covariance matrix estimate becomes singular during EM. The Bayesian regularization method proposed by Fraley and Raftery (2007) and implemented in **mclust** as described in Section 7.2 addresses this problem.

The top three models are also listed at the bottom of the table. A summary of the top models according to the BIC criterion can also be obtained as

```
summary(mod$BIC, k = 5)
## Best BIC values:
##                 VVI,3    VVI,4    VVV,3    VVI,5    VVI,6
```

```
## BIC      -4777.9 -4796.773 -4809.761 -4810.928 -4854.644
## BIC diff     0.0   -18.866   -31.854   -33.021   -76.737
```

where we have specified that the the top five GMMs should be shown through
the optional argument k = 5. This summary is useful because it provides the
BIC differences with respect to the best model, namely, the one with the
largest BIC. For the interpretation of BIC, see the discussion in Section 2.3.1.

The BIC values can be displayed graphically using

```
plot(mod, what = "BIC",
     legendArgs = list("bottomright", ncol = 5))
```

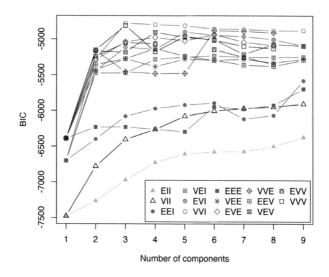

FIGURE 3.7: BIC traces for the GMMs estimated for the thyroid data.

where we included the optional argument legendArgs to control the placement
and format of the legend. The resulting plot of BIC traces is shown in Figure 3.7.

The model selected is (VVI,3), a three-component Gaussian mixture with
diagonal covariance structure, in which both the volume and shape vary
between clusters. We can obtain a full summary of the model estimates as
follows:

```
summary(mod, parameters = TRUE)
## ------------------------------------------------------
## Gaussian finite mixture model fitted by EM algorithm
## ------------------------------------------------------
##
## Mclust VVI (diagonal, varying volume and shape) model with 3
```

```
## components:
##
##  log-likelihood   n df     BIC     ICL
##           -2303 215 32 -4777.9 -4784.5
##
## Clustering table:
##   1   2   3
##  35 152  28
##
## Mixing probabilities:
##       1       2       3
## 0.16310 0.70747 0.12943
##
## Means:
##          [,1]     [,2]     [,3]
## RT3U 95.536865 110.3462 123.2078
## T4   17.678383   9.0880   3.7994
## T3    4.262681   1.7218   1.0576
## TSH   0.970155   1.3051  13.8951
## DTSH -0.017415   2.4960  18.8221
##
## Variances:
## [,,1]
##         RT3U     T4     T3     TSH     DTSH
## RT3U 343.92  0.000 0.0000 0.00000 0.000000
## T4     0.00 17.497 0.0000 0.00000 0.000000
## T3     0.00  0.000 4.9219 0.00000 0.000000
## TSH    0.00  0.000 0.0000 0.15325 0.000000
## DTSH   0.00  0.000 0.0000 0.00000 0.071044
## [,,2]
##         RT3U     T4      T3     TSH   DTSH
## RT3U 66.372 0.0000 0.00000 0.00000 0.0000
## T4    0.000 4.8234 0.00000 0.00000 0.0000
## T3    0.000 0.0000 0.23278 0.00000 0.0000
## TSH   0.000 0.0000 0.00000 0.21394 0.0000
## DTSH  0.000 0.0000 0.00000 0.00000 3.1892
## [,,3]
##        RT3U    T4      T3 TSH    DTSH
## RT3U 95.27 0.000 0.00000   0    0.00
## T4    0.00 4.278 0.00000   0    0.00
## T3    0.00 0.000 0.27698   0    0.00
## TSH   0.00 0.000 0.00000 147    0.00
## DTSH  0.00 0.000 0.00000   0  231.29
```

The clustering implied by the selected model can then be shown graphically:

```
plot(mod, what = "classification")
```

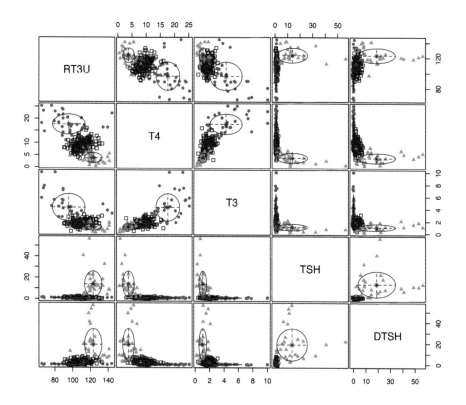

FIGURE 3.8: Scatterplot matrix for the thyroid data with points marked according to the GMM (VVI,3) clustering, and ellipses corresponding to projections of the estimated cluster covariances.

(see Figure 3.8). The confusion matrix for the clustering and the corresponding adjusted Rand index are obtained as:

```
table(Class, Cluster = mod$classification)
##          Cluster
## Class     1   2   3
##   Hypo    0   4  26
##   Normal  1 147   2
##   Hyper  34   1   0
adjustedRandIndex(Class, mod$classification)
## [1] 0.87715
```

The estimated clustering model turns out to be quite accurate in recovering the true thyroid diagnosis.

A plot of the posterior conditional probabilities ordered by cluster can be obtained using the following code:

```
z  <- mod$z                  # posterior conditional probabilities
cl <- mod$classification  # MAP clustering
G  <- mod$G                  # number of clusters
sclass <- 10 # class separation
sedge <- 3    # edge spacing
L <- nrow(z) + G*(sclass+2*sedge)
plot(1:L, runif(L), ylim = c(0, 1), type = "n", axes = FALSE,
     ylab = "Posterior conditional probabilities", xlab = "")
axis(2)
col <- mclust.options("classPlotColors")
l <- sclass
for (k in 1:G)
{
  i <- which(cl == k)
  ord <- i[order(z[i, k], decreasing = TRUE)]
  for (j in 1:G)
     points((l+sedge)+1:length(i), z[ord, j],
            pch = as.character(j), col = col[j])
  rect(l, 0, l+2*sedge+length(i), 1,
       border = col[k], col = col[k], lwd = 2, density = 0)
  l <- l + 2*sedge + length(i) + sclass
}
```

This yields Figure 3.9, in which a panel of posterior conditional probabilities of class membership is shown for each of the three clusters determined by the MAP classification. The plot shows that observations in the third cluster have the largest probabilities of membership, while there is some uncertainty in membership among the first two clusters.

EXAMPLE 3.3: Clustering Italian wines data

Forina et al. (1986) reported data on 178 wines grown in the same region in Italy but derived from three different cultivars (Barbera, Barolo, Grignolino). For each wine, 13 measurements of chemical and physical properties were made as reported in the table below. This data set is from the UCI Machine Learning Data Repository (Dua and Graff (2017) — see http://archive.ics.uci.edu/ml /datasets/Wine) and is available in the **gclus** package (Hurley, 2019).

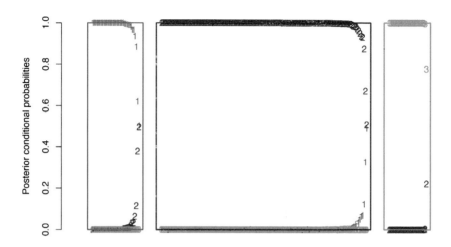

Posterior conditional probabilities

FIGURE 3.9: Estimated posterior conditional probabilities of class membership for each of the three clusters determined by the MAP classification of observations in the thyroid data. The panels correspond to the different clusters.

Variable	Description
Class	Wine classes: 1) Barolo; 2) Grignolino; 3) Barbera
Alcohol	Alcohol
Malic	Malic acid
Ash	Ash
Alcalinity	Alcalinity of ash
Magnesium	Magnesium
Phenols	Total phenols
Flavanoids	Flavanoids
Nonflavanoid	Nonflavanoid phenols
Proanthocyanins	Proanthocyanins
Intensity	Color intensity
Hue	Hue
OD280	OD280/OD315 of diluted wines
Proline	Proline

```
data("wine", package = "gclus")
Class <- factor(wine$Class, levels = 1:3,
                labels = c("Barolo", "Grignolino", "Barbera"))
X <- data.matrix(wine[, -1])
```

For a description of the data, see `help("wine", package = "gclus")`.

We fit the data with `Mclust()` for all 14 models, and then summarize the BIC for the top 3 models:

```
mod <- Mclust(X)
summary(mod$BIC, k = 3)
## Best BIC values:
##              VVE,3      EVE,4      VVE,4
## BIC      -6849.4 -6873.616 -6885.472
## BIC diff    0.0   -24.225   -36.081
```

The BIC traces are then plotted (see Figure 3.10):

```
plot(mod, what = "BIC",
     ylim = range(mod$BIC[, -(1:2)], na.rm = TRUE),
     legendArgs = list(x = "bottomleft"))
```

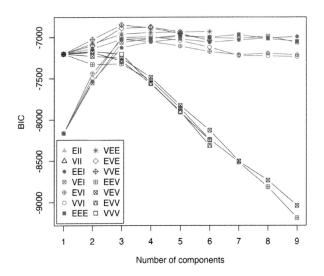

FIGURE 3.10: BIC plot for models fitted to the `wine` data.

Note that, in this last plot, we adjusted the range of the vertical axis so as to remove models with lower BIC values. The model selected by BIC is a three-component Gaussian mixture with covariances having different volumes and shapes, but the same orientation (VVE). This is a flexible but relatively parsimonious model. A summary of the selected model is then obtained:

```
summary(mod)
## -----------------------------------------------------
```

```
## Gaussian finite mixture model fitted by EM algorithm
## -------------------------------------------------------
##
## Mclust VVE (ellipsoidal, equal orientation) model with 3 components:
##
## log-likelihood    n  df     BIC     ICL
##        -3015.3  178 158 -6849.4 -6850.7
##
## Clustering table:
##  1  2  3
## 59 69 50
table(Class, mod$classification)
##
## Class          1  2  3
##    Barolo     59  0  0
##    Grignolino  0 69  2
##    Barbera     0  0 48
adjustedRandIndex(Class, mod$classification)
## [1] 0.96674
```

The fitted model provides an accurate recovery of the true classes with a high ARI.

Displaying the results is difficult because of the high dimensionality of the data. Below we give a simple method for choosing a subset of variables for a matrix of scatterplots. More sophisticated approaches to choosing projections that reveal clustering are discussed in Section 6.4. First we plot the estimated cluster means, normalized to the $[0, 1]$ range, using the heatmap function available in base R:

```
norm01 <- function(x) (x - min(x))/(max(x) - min(x))
M <- apply(t(mod$parameters$mean), 2, norm01)
heatmap(M, Rowv = NA, scale = "none", margins = c(8, 2),
        labRow = paste("Cluster", 1:mod$G), cexRow = 1.2)
```

The resulting heatmap (see Figure 3.11) shows which features have the most different cluster means. Based on this, we select a subset of three variables (Alcohol, Malic, and Flavanoids) for the dimens argument specifying coordinate projections for multivariate data in the plot() function call:

```
plot(mod, what = "classification", dimens = c(1, 2, 7))
```

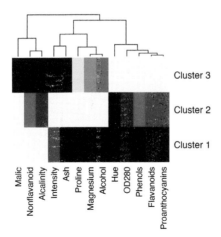

FIGURE 3.11: Heatmap of normalized cluster means for the clustering model fitted to the wine data.

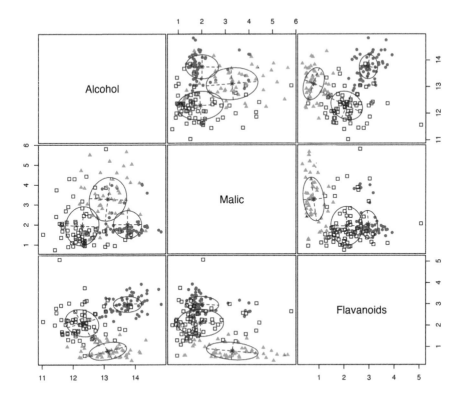

FIGURE 3.12: Coordinate projection plot of selected features showing the clusters for the wine data.

3.3   Model Selection

3.3.1   BIC

In the **mclust** package, BIC is used by default for model selection. The function mclustBIC(), and indirectly Mclust(), computes a matrix of BIC values for models chosen from Table 3.1, and numbers of mixture components as specified.

EXAMPLE 3.4: BIC model selection for the Old Faithful data

As an illustration, consider the bivariate faithful dataset (Härdle, 1991), included in the base R package **datasets**. "Old Faithful" is a geyser located in Yellowstone National Park, Wyoming, USA. The faithful dataset provides the duration of Old Faithful's eruptions (eruptions, time in minutes) and the waiting times between successive eruptions (waiting, time in minutes). A scatterplot of the data is shown in panel Figure 3.13, obtained using the code:

```
data("faithful", package = "datasets")
plot(faithful)
```

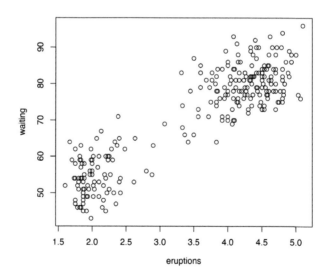

FIGURE 3.13: Scatterplot of the Old Faithful data available in the faithful dataset.

The following code can be used to compute the BIC values and plot the corresponding BIC traces:

```
BIC <- mclustBIC(faithful)
BIC
## Bayesian Information Criterion (BIC):
##        EII      VII      EEI      VEI      EVI      VVI      EEE      VEE
## 1 -4024.7 -4024.7 -3055.8 -3055.8 -3055.8 -3055.8 -2607.6 -2607.6
## 2 -3453.0 -3458.3 -2354.6 -2350.6 -2352.6 -2346.1 -2325.2 -2323.0
## 3 -3377.7 -3336.6 -2323.0 -2332.7 -2332.2 -2342.4 -2314.3 -2322.1
## 4 -3230.3 -3242.8 -2323.7 -2331.3 -2334.7 -2343.5 -2331.2 -2340.2
## 5 -3149.4 -3129.1 -2327.1 -2350.2 -2347.6 -2351.0 -2360.7 -2347.3
## 6 -3081.4 -3038.2 -2338.2 -2360.6 -2357.7 -2373.5 -2347.4 -2372.3
## 7 -2990.4 -2973.4 -2356.5 -2368.5 -2372.9 -2394.7 -2369.3 -2371.2
## 8 -2978.1 -2935.1 -2364.1 -2384.7 -2389.1 -2413.7 -2376.1 -2390.4
## 9 -2953.4 -2919.4 -2372.8 -2398.2 -2407.2 -2432.7 -2389.6 -2406.7
##        EVE      VVE      EEV      VEV      EVV      VVV
## 1 -2607.6 -2607.6 -2607.6 -2607.6 -2607.6 -2607.6
## 2 -2324.3 -2320.4 -2329.1 -2325.4 -2327.6 -2322.2
## 3 -2342.3 -2336.3 -2325.3 -2329.6 -2340.0 -2349.7
## 4 -2361.8 -2362.5 -2351.5 -2361.1 -2344.7 -2351.5
## 5 -2351.8 -2368.9 -2356.9 -2368.1 -2364.9 -2379.4
## 6 -2366.5 -2386.5 -2366.1 -2386.3 -2384.1 -2387.0
## 7 -2379.8 -2402.2 -2379.1 -2401.3 -2398.7 -2412.4
## 8 -2403.9 -2426.0 -2393.0 -2425.4 -2415.0 -2442.0
## 9 -2414.1 -2448.2 -2407.5 -2446.7 -2438.9 -2460.4
##
## Top 3 models based on the BIC criterion:
##     EEE,3    VVE,2    VEE,3
## -2314.3 -2320.4 -2322.1
plot(BIC)
```

In this case the BIC supports a three-component GMM with common full covariance matrix (EEE model). Note that, as for the thyroid disease data example, there are missing BIC values (denoted by NA) corresponding to models that could not be estimated as initialized.

By default, mclustBIC computes results for up to 9 components and all available models in Table 3.1. Optional arguments can be provided to mclustBIC() allowing fine tuning, such as G for the number of components, and modelNames for specifying the model covariance parameterizations (see Table 3.1 and help("mclustModelNames") for a description of available model names).

Another optional argument, called x, can be used to provide the output from a previous call to mclustBIC(). This is useful if the model space needs to be enlarged by fitting more models (by increasing the number of mixture components and/or specifying additional covariance parameterizations) without the need to recompute the BIC values for those models already fitted. BIC values already available can be provided analogously to Mclust as follows:

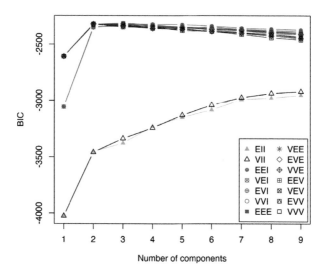

FIGURE 3.14: BIC traces for the GMMs estimated for the faithful dataset.

```
mod1 <- Mclust(faithful, x = BIC)
```

3.3.2   ICL

Besides the BIC criterion for model selection, the *integrated complete-data likelihood* (ICL) criterion, described in Section 2.3.1, is available in **mclust** through the function mclustICL(). It is invoked similarly to mclustBIC(), and it allows us to select both the model (among those in Table 3.1) and the number of mixture components that maximize the ICL criterion.

EXAMPLE 3.5: ICL model selection for the Old Faithful data

Considering again the data in Example 3.4, ICL values are computed as follows:

```
ICL <- mclustICL(faithful)
ICL
## Integrated Complete-data Likelihood (ICL) criterion:
##        EII       VII      EEI      VEI      EVI      VVI      EEE      VEE
## 1  -4024.7  -4024.7  -3055.8  -3055.8  -3055.8  -3055.8  -2607.6  -2607.6
## 2  -3455.8  -3460.9  -2356.3  -2350.7  -2353.3  -2346.2  -2326.7  -2323.4
## 3  -3422.8  -3360.3  -2359.5  -2377.3  -2367.5  -2387.7  -2357.8  -2376.5
## 4  -3265.8  -3272.5  -2372.0  -2413.4  -2402.2  -2436.3  -2468.3  -2452.7
## 5  -3190.7  -3151.9  -2394.0  -2486.7  -2412.4  -2445.8  -2478.2  -2472.0
```

```
## 6 -3117.4 -3061.3 -2423.0 -2486.8 -2446.9 -2472.6 -2456.2 -2503.9
## 7 -3022.3 -2995.8 -2476.2 -2519.8 -2446.7 -2496.7 -2464.3 -2466.8
## 8 -3007.4 -2953.7 -2488.5 -2513.5 -2492.3 -2509.7 -2502.2 -2479.8
## 9 -2989.1 -2933.1 -2499.9 -2540.4 -2515.0 -2528.6 -2547.1 -2499.9
##          EVE      VVE      EEV      VEV      EVV      VVV
## 1 -2607.6 -2607.6 -2607.6 -2607.6 -2607.6 -2607.6
## 2 -2325.8 -2320.8 -2330.0 -2325.7 -2328.2 -2322.7
## 3 -2412.0 -2427.0 -2372.4 -2405.3 -2380.3 -2385.2
## 4 -2459.4 -2440.3 -2414.2 -2419.9 -2385.8 -2407.6
## 5 -2444.3 -2478.6 -2431.1 -2490.2 -2423.2 -2474.5
## 6 -2504.8 -2489.1 -2449.6 -2481.4 -2483.8 -2491.6
## 7 -2499.3 -2496.3 -2465.7 -2506.8 -2490.1 -2519.5
## 8 -2526.0 -2516.6 -2489.4 -2539.8 -2497.8 -2556.1
## 9 -2545.7 -2541.7 -2542.9 -2566.7 -2528.6 -2587.2
##
## Top 3 models based on the ICL criterion:
##    VVE,2    VVV,2    VEE,2
## -2320.8 -2322.7 -2323.4
```

A trace of ICL values is plotted and shown in Figure 3.15:

```
plot(ICL)
```

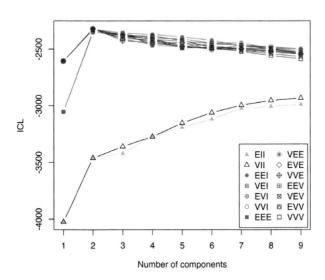

FIGURE 3.15: ICL traces for the GMMs estimated for the `faithful` dataset.

Note that, in this case, ICL selects two components, one less than the number selected by BIC. This is reasonable, since ICL penalizes overlapping components more heavily.

The model selected by ICL can be fitted using

```
mod2 <- Mclust(faithful, G = 2, modelNames = "VVE")
```

Optional arguments can also be provided in the Mclust() function call, analogous to those described above for mclustBIC().

The following code can be used to draw scatterplots of the data with points marked according to the classification obtained by the "best" estimated model selected, respectively, by the BIC and ICL model selection criteria (see Figure 3.16):

```
plot(mod1, what = "classification", fillEllipses = TRUE)
plot(mod2, what = "classification", fillEllipses = TRUE)
```

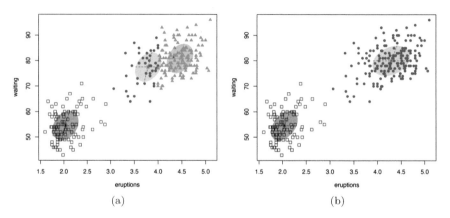

FIGURE 3.16: Scatterplots for the faithful dataset with points and ellipses corresponding to the classification from the best estimated GMMs selected by BIC (a) and ICL (b).

3.3.3   Bootstrap Likelihood Ratio Testing

The bootstrap procedure discussed in Section 2.3.2 for selecting the number of mixture components is implemented in the mclustBootstrapLRT() function.

EXAMPLE 3.6: Bootstrap LRT for the Old Faithful data

Using the data of Example 3.4, a bootstrap LR test is obtained by specifying the input data and the name of the model to test:

```
LRT <- mclustBootstrapLRT(faithful, modelName = "VVV")
LRT
## ------------------------------------------------------------
## Bootstrap sequential LRT for the number of mixture components
## ------------------------------------------------------------
## Model        = VVV
## Replications = 999
##                 LRTS bootstrap p-value
## 1 vs 2    319.0654              0.001
## 2 vs 3      6.1305              0.549
```

The sequential bootstrap procedure terminates when a test is not significant as specified by the argument level (which is set to 0.05 by default). There is also the option to specify the maximum number of mixture components to test via the argument maxG. The number of bootstrap resamples can be set with the optional argument nboot (nboot = 999 is the default).

In the example above, the bootstrap p-values clearly indicate the presence of two clusters. Note that models fitted to the original data are estimated via the EM algorithm initialized with unconstrained model-based agglomerative hierarchical clustering (the default). Then, during the bootstrap procedure, models under the null and the alternative hypotheses are fitted to bootstrap samples using again the EM algorithm. However, in this case the algorithm starts with the E-step initialized with the estimated parameters obtained for the original data.

The bootstrap distributions of the LRTS can be shown graphically (see Figure 3.17) using the associated plot method:

```
plot(LRT, G = 1)
plot(LRT, G = 2)
```

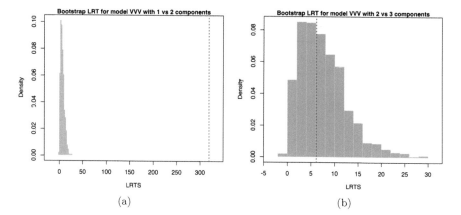

FIGURE 3.17: Histograms of LRTS bootstrap distributions for testing the number of mixture components in the faithful data assuming the VVV model. The dotted vertical lines refer to the sample values of LRTS.

3.4   Resampling-Based Inference in mclust

Section 2.4 describes some resampling-based approaches to inference in GMMs, namely the nonparametric bootstrap, the parametric bootstrap, and the weighted likelihood bootstrap. These, together with the *jackknife* (Efron, 1979, 1982), are implemented in the MclustBootstrap() function available in **mclust**.

EXAMPLE 3.7: Resampling-based inference for Gaussian mixtures on the hemophilia data

Consider the hemophilia dataset (Habbema et al., 1974) available in the package **rrcov** (Todorov, 2022), which contains two measured variables on 75 women belonging to two groups: 30 of them are non-carriers (normal group) and 45 are known hemophilia A carriers (obligatory carriers).

```
data("hemophilia", package = "rrcov")
X <- hemophilia[, 1:2]
Class <- as.factor(hemophilia$gr)
clp <- clPairs(X, Class, symbols = c(16, 0), colors = "black")
clPairsLegend(0.8, 0.2, class = clp$class, col = clp$col, pch = clp$pch)
```

The last command plots the known classification of the observed data (see Figure 3.18a).

By analogy with the analysis of Basford et al. (1997, Example II, Section 5), we fitted a two-component GMM with unconstrained covariance matrices:

```
mod <- Mclust(X, G = 2, modelName = "VVV")
summary(mod, parameters = TRUE)
## ---------------------------------------------------------
## Gaussian finite mixture model fitted by EM algorithm
## ---------------------------------------------------------
##
## Mclust VVV (ellipsoidal, varying volume, shape, and orientation) model
## with 2 components:
##
##   log-likelihood  n df     BIC     ICL
##           77.029 75 11  106.56 92.855
##
## Clustering table:
##   1  2
## 39 36
##
## Mixing probabilities:
##        1        2
## 0.51015 0.48985
##
## Means:
##                   [,1]      [,2]
## AHFactivity -0.116124 -0.366387
## AHFantigen  -0.024573 -0.045323
##
## Variances:
## [,,1]
##             AHFactivity AHFantigen
## AHFactivity   0.0113599  0.0065959
## AHFantigen    0.0065959  0.0123872
## [,,2]
##             AHFactivity AHFantigen
## AHFactivity    0.015874   0.015050
## AHFantigen     0.015050   0.032341
```

Note that, in the summary() function call, we set parameters = TRUE to retrieve the estimated parameters. The clustering structure identified is shown in Figure 3.18b and can be obtained as follows:

```
plot(mod, what = "classification", fillEllipses = TRUE)
```

A function, MclustBootstrap(), is available for bootstrap inference for GMMs. This function requires the user to input an object as returned by a call to Mclust(). Optionally, the number of bootstrap resamples (nboot) and the type

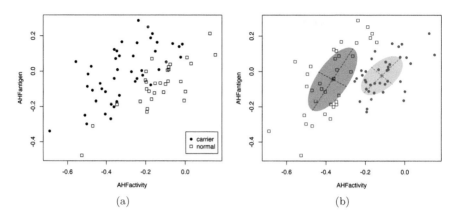

FIGURE 3.18: Scatterplots for the hemophilia data displaying the true class membership (a) and the classification obtained from the fit of a GMM (b).

of bootstrap resampling to use (type), can also be specified. By default, nboot = 999 and type = "bs", with the latter specifying the nonparametric bootstrap. Thus, a simple call for computing the bootstrap distribution of the GMM parameters is the following:

```
boot <- MclustBootstrap(mod, nboot = 999, type = "bs")
```

Note that, although we have included the arguments nboot and type, they could have been omitted in this instance since they are both set at their defaults.

Results of the function MclustBootstrap() are available through the summary() method, which by default returns the standard errors of the GMM parameters:

```
summary(boot, what = "se")
## ---------------------------------------------------------------
## Resampling standard errors
## ---------------------------------------------------------------
## Model                       = VVV
## Num. of mixture components = 2
## Replications                = 999
## Type                        = nonparametric bootstrap
##
## Mixing probabilities:
##       1       2
## 0.12254 0.12254
##
## Means:
```

```
##                        1        2
## AHFactivity 0.038593 0.04122
## AHFantigen  0.031411 0.06492
##
## Variances:
## [,,1]
##             AHFactivity AHFantigen
## AHFactivity   0.0068325  0.0046163
## AHFantigen    0.0046163  0.0031646
## [,,2]
##             AHFactivity AHFantigen
## AHFactivity   0.0057492  0.0057642
## AHFantigen    0.0057642  0.0098679
```

Bootstrap percentile confidence intervals, described in Section 2.4, can also be obtained by specifying what = "ci" in the summary() call. The confidence level of the intervals can also be specified (by default, conf.level = 0.95). For instance:

```
summary(boot, what = "ci", conf.level = 0.9)
## ------------------------------------------------------------
## Resampling confidence intervals
## ------------------------------------------------------------
## Model                     = VVV
## Num. of mixture components = 2
## Replications              = 999
## Type                      = nonparametric bootstrap
## Confidence level          = 0.9
##
## Mixing probabilities:
##           1       2
## 5%  0.38429 0.21214
## 95% 0.78786 0.61571
##
## Means:
## [,,1]
##     AHFactivity AHFantigen
## 5%    -0.211979  -0.079614
## 95%   -0.086514   0.018313
## [,,2]
##     AHFactivity AHFantigen
## 5%    -0.44176  -0.143275
## 95%   -0.31241   0.098511
##
## Variances:
```

```
## [,,1]
##      AHFactivity AHFantigen
## 5%     0.0057382  0.0076932
## 95%    0.0283428  0.0179437
## [,,2]
##      AHFactivity AHFantigen
## 5%     0.0048406  0.0088428
## 95%    0.0234544  0.0426075
```

The bootstrap distribution of the parameters can also be represented graphically.
For instance, the following code produces a plot of the bootstrap distribution
for the mixing proportions:

```
par(mfrow = c(1, 2))
plot(boot, what = "pro")
```

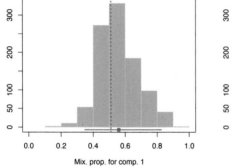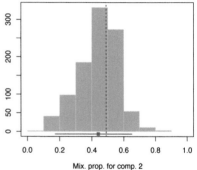

FIGURE 3.19: Bootstrap distribution for the mixture proportions of the GMM
fitted to the hemophilia data. The vertical dotted lines indicate the MLEs, the
bottom line indicates the percentile confidence intervals, and the square shows
the center of the bootstrap distribution.

and for the component means:

```
par(mfcol = c(2, 2))
plot(boot, what = "mean")
```

The resulting plots are shown in Figures 3.19 and 3.20.

As mentioned, with the function MclustBootstrap(), it is also possible
to choose the parametric bootstrap (type = "pb"), the weighted likelihood
bootstrap (type = "wlbs"), or the jackknife (type = "jk"). For instance, in our
data example the weighted likelihood bootstrap can be obtained as follows:

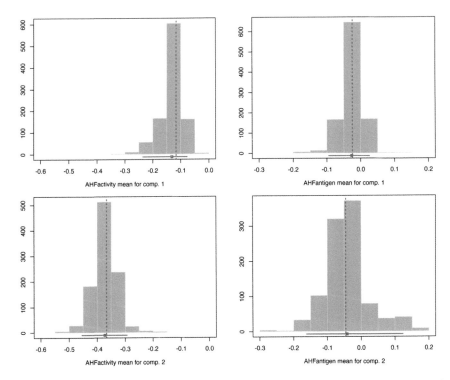

FIGURE 3.20: Bootstrap distribution for the mixture component means of the GMM fitted to the hemophilia data. The vertical dotted lines indicate the MLEs, the bottom line indicates the percentile confidence intervals, and the square shows the center of the bootstrap distribution.

```
wlboot <- MclustBootstrap(mod, nboot = 999, type = "wlbs")
summary(wlboot, what = "se")
## ----------------------------------------------------------
## Resampling standard errors
## ----------------------------------------------------------
## Model                     = VVV
## Num. of mixture components = 2
## Replications              = 999
## Type                      = weighted likelihood bootstrap
##
## Mixing probabilities:
##       1       2
## 0.12669 0.12669
##
## Means:
```

```
##                       1        2
## AHFactivity 0.039016 0.041494
## AHFantigen  0.030487 0.063628
##
## Variances:
## [,,1]
##              AHFactivity AHFantigen
## AHFactivity   0.0067704  0.0045065
## AHFantigen    0.0045065  0.0031056
## [,,2]
##              AHFactivity AHFantigen
## AHFactivity   0.0057317  0.0059695
## AHFantigen    0.0059695  0.0091836
```

In this case the differences between the nonparametric and the weighted likelihood bootstrap are negligible. We can summarize the inference for the GMM component means obtained under the two approaches graphically, showing the bootstrap percentile confidence intervals for each variable and component (see Figure 3.21):

```
boot.ci <- summary(boot, what = "ci")
wlboot.ci <- summary(wlboot, what = "ci")
for (j in 1:mod$d)
{
  plot(1:mod$G, mod$parameters$mean[j, ], col = 1:mod$G, pch = 15,
       ylab = colnames(X)[j], xlab = "Mixture component",
       ylim = range(boot.ci$mean, wlboot.ci$mean),
       xlim = c(.5, mod$G+.5), xaxt = "n")
  points(1:mod$G+0.2, mod$parameters$mean[j, ], col = 1:mod$G, pch = 15)
  axis(side = 1, at = 1:mod$G)
  with(boot.ci,
       errorBars(1:G, mean[1, j, ], mean[2, j, ], col = 1:G))
  with(wlboot.ci,
       errorBars(1:G+0.2, mean[1, j, ], mean[2, j, ], col = 1:G, lty = 2))
}
```

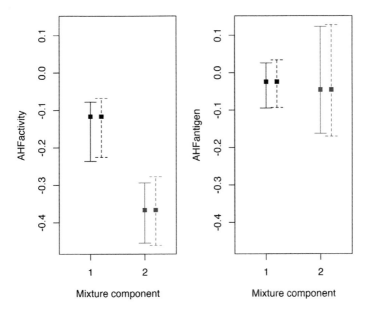

FIGURE 3.21: Bootstrap percentile intervals for the means of the GMM fitted to the hemophilia data. Solid lines refer to the nonparametric bootstrap, dashed lines to the weighted likelihood bootstrap.

3.5 Clustering Univariate Data

For clustering univariate data, the quantiles of the empirical distribution are used to initialize the EM algorithm by default, rather than hierarchical clustering. There are only two possible models, E for equal variance across components and V allowing the variance to vary across the components.

EXAMPLE 3.8: Univariate clustering of annual precipitation in US cities

As an example of univariate clustering, consider the precip dataset (McNeil, 1977) included in the package **datasets** available in the base R distribution:

```
data("precip", package = "datasets")
dotchart(sort(precip), cex = 0.6, pch = 19,
         xlab = "Average annual rainfall (in inches)")
```

The average amount of rainfall (in inches) for each of 70 US cities is shown as a dot chart in Figure 3.22. A clustering model is then obtained:

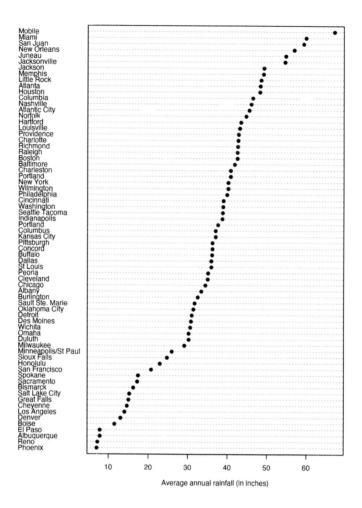

FIGURE 3.22: Dot chart of average annual rainfall (in inches) for 70 US cities.

```
mod <- Mclust(precip)
summary(mod, parameters = TRUE)
## -------------------------------------------------------
## Gaussian finite mixture model fitted by EM algorithm
## -------------------------------------------------------
##
## Mclust V (univariate, unequal variance) model with 2 components:
##
## log-likelihood  n df     BIC     ICL
##        -275.47 70  5 -572.19 -575.21
```

```
##
## Clustering table:
## 1  2
## 13 57
##
## Mixing probabilities:
##       1       2
## 0.18137 0.81863
##
## Means:
##       1       2
## 12.793 39.780
##
## Variances:
##       1       2
## 16.813 90.394
plot(mod, what = "BIC", legendArgs = list(x = "bottomleft"))
```

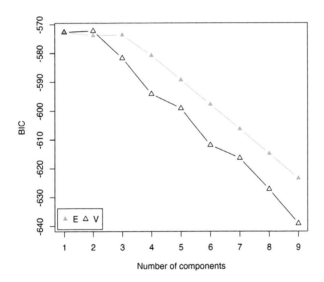

FIGURE 3.23: BIC traces for GMMs fitted to the precip data.

The selected model is a two-component Gaussian mixture with different variances. The first cluster, which includes 13 cities, is characterized by both smaller means and variances compared to the second cluster. The clustering partition and the corresponding classification uncertainties can be displayed using the following plot commands (see Figure 3.24):

```
plot(mod, what = "classification")
plot(mod, what = "uncertainty")
```

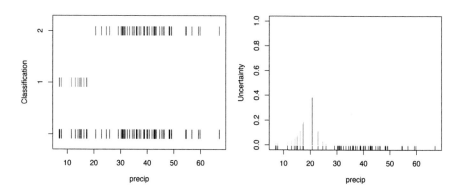

FIGURE 3.24: Classification (left) and uncertainty (right) plots for the precip data.

The final clustering is shown in a dot chart with observations grouped by clusters (see Figure 3.25):

```
x <- data.frame(precip, clusters = mod$classification)
rownames(x) <- make.unique(names(precip)) # correct duplicated names
x <- x[order(x$precip), ]
dotchart(x$precip, labels = rownames(x),
         # groups = factor(x$clusters, labels = c("Cluster 1", "Cluster 2")),
         groups = factor(x$clusters, levels = 2:1,
                         labels = c("Cluster 2", "Cluster 1")),
         cex = 0.6, pch = 19,
         color = mclust.options("classPlotColors")[x$clusters],
         xlab = "Average annual rainfall (in inches)")
```

Each of the graphs shown indicates that the two groups of cities are separated by about 20 inches of annual rainfall.

Further graphical capabilities for univariate data are discussed in Section 6.1.

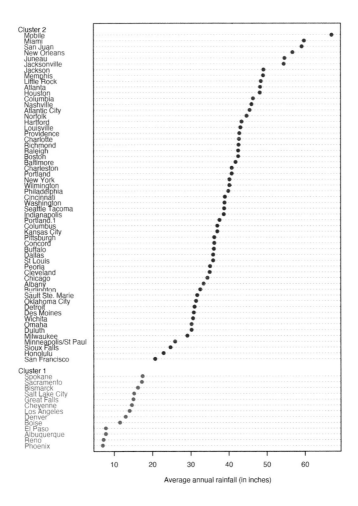

FIGURE 3.25: Dot chart of average annual rainfall (in inches) for 70 US cities grouped by the estimated clustering partitions.

3.6   Model-Based Agglomerative Hierarchical Clustering

Model-based agglomerative hierarchical clustering (MBAHC; Banfield and Raftery, 1993) adopts the general ideas of classical agglomerative hierarchical clustering, with the aim of obtaining a hierarchy of clusters of n objects. Given no preliminary information with regard to grouping, agglomerative hierarchical clustering proceeds from n clusters each containing a single observation, to one cluster containing all n observations, by successively merging observations and clusters. In the traditional approach, two groups are merged if they are closest according to a particular distance metric and type of linkage, while in MBAHC, the two groups are merged that optimize a regularized criterion (Fraley, 1998) derived from the classification likelihood over all possible merge pairs.

Given n observations $(\boldsymbol{x}_1, \ldots, \boldsymbol{x}_n)$, let $(\ell_1, \ldots, \ell_n)^\top$ denote the classification labels, so that $\ell_i = k$ if $\boldsymbol{x}_i$ is assigned to cluster k. The unknown parameters are obtained by maximizing the classification likelihood:

$$L_{CL}(\boldsymbol{\theta}_1, \ldots, \boldsymbol{\theta}_G, \ell_1, \ldots, \ell_n; \boldsymbol{x}_1, \ldots, \boldsymbol{x}_n) = \prod_{i=1}^{n} f_{\ell_i}(\boldsymbol{x}_i; \boldsymbol{\theta}_{\ell_i}).$$

Assuming a multivariate Gaussian distribution for the data, so that $f_{\ell_i}(\boldsymbol{x}_i; \boldsymbol{\theta}_{\ell_i}) = \phi(\boldsymbol{x}_i; \boldsymbol{\mu}_k, \boldsymbol{\Sigma}_k)$ if $\ell_i = k$, the classification likelihood is

$$L_{CL}(\boldsymbol{\mu}_1, \ldots, \boldsymbol{\mu}_G, \boldsymbol{\Sigma}_1, \ldots, \boldsymbol{\Sigma}_G, \ell_1, \ldots, \ell_n; \boldsymbol{x}_1, \ldots, \boldsymbol{x}_n) = \prod_{k=1}^{G} \prod_{i \in \mathcal{I}_k} \phi(\boldsymbol{x}_i; \boldsymbol{\mu}_k, \boldsymbol{\Sigma}_k),$$

where $\mathcal{I}_k = \{i : \ell_i = k\}$ is the set of indices corresponding to the observations that belong to cluster k. After replacing the unknown mean vectors $\boldsymbol{\mu}_k$ by their MLEs, $\widehat{\boldsymbol{\mu}}_k = \bar{\boldsymbol{x}}_k$, the profile (or concentrated) log-likelihood is given by

$$\log L_{CL}(\boldsymbol{\Sigma}_1, \ldots, \boldsymbol{\Sigma}_G, \ell_1, \ldots, \ell_n; \boldsymbol{x}_1, \ldots, \boldsymbol{x}_n) =$$
$$- \frac{1}{2} nd \log(2\pi) - \frac{1}{2} \sum_{k=1}^{G} \{ \mathrm{tr}(\boldsymbol{W}_k \boldsymbol{\Sigma}_k^{-1}) + n_k \log |\boldsymbol{\Sigma}_k| \}, \quad (3.1)$$

where $\boldsymbol{W}_k = \sum_{i \in \mathcal{I}_k} (\boldsymbol{x}_i - \bar{\boldsymbol{x}}_k)(\boldsymbol{x}_i - \bar{\boldsymbol{x}}_k)^\top$ is the sample cross-product matrix for the kth cluster, and n_k is the number of observations in cluster k.

By imposing different covariance structures on $\boldsymbol{\Sigma}_k$, different merging criteria can be derived (see Fraley, 1998, Table 1). For instance, assuming a common isotropic cluster-covariance, $\boldsymbol{\Sigma}_k = \sigma^2 \boldsymbol{I}$, the log-likelihood in (3.1) is maximized by minimizing the criterion

$$\mathrm{tr}\left(\sum_{k=1}^{G} \boldsymbol{W}_k \right),$$

the well-known sum of squares criterion of Ward (1963). If the covariances are allowed to be different across clusters, then the criterion to be minimized at each stage of the hierarchical algorithm is

$$\sum_{k=1}^{G} n_k \log \left| \frac{\boldsymbol{W}_k}{n_k} \right|.$$

Methods for regularizing these criteria and efficient numerical algorithms for their minimization have been discussed by Fraley (1998). Table 3.1 lists the models for hierarchical clustering available in **mclust**.

At each stage of MBAHC, a pair of groups is merged that minimizes a regularized merge criterion over all possible merge pairs (Fraley, 1998). Regularization is necessary because cluster contributions to the merge criterion (and the classification likelihood) are undefined for singletons, as well as for other combinations of observations resulting in a singular or near-singular sample covariance.

Traditional hierarchical clustering procedures, as implemented, for example, in the base R function hclust(), are typically represented graphically with a *dendrogram* that shows the links between objects using a tree diagram, usually with the root at the top and leaves at the bottom. The height of the branches reflects the order in which the clusters are joined, and, to some extent, it also reflects the distances between clusters (see, for example Hastie et al., 2009, Section 14.3.12).

Because the regularized merge criterion may either increase or decrease from stage to stage, the association of a positive distance with merges in traditional hierarchical clustering does not apply to MBAHC. However, it is still possible to draw a dendrogram in the which the tree height increases as the number of groups decreases, so that the root is at the top of the hierarchy. Moreover, the classification likelihood often increases in the later stages of merging (small numbers of clusters), in which case the corresponding dendrogram can be drawn.

mclust provides the function hc() to perform model-based agglomerative hierarchical clustering. It takes the data matrix to be clustered as the main argument. The principal optional arguments are as follows:

modelName, for selecting the model to fit among the four models available, which, following the nomenclature in Table 2.1, are denoted as EII, VII, EEE, and VVV (default).

partition, for providing an initial partition from which to start the agglomeration. The default is to start with partitioning into unique observations.

use, for specifying the data transformation to apply before performing hierarchical clustering. By default use = "VARS", so the variables are expressed in the original scale. Other possible values, including the default for initialization of the EM algorithm in model-based clustering, are described in Section 3.7.

EXAMPLE 3.9: Model-based hierarchical cluastering of European unemployment data

Consider the EuroUnemployment dataset which provides unemployment rates for 31 European countries for the year 2014. The considered rates are shown in the table below.

Variable	Description
TUR	Total unemployment rate, defined as the percentage of unemployed persons aged 15–74 in the economically active population.
YUR	Youth unemployment rate, defined as the percentage of unemployed persons aged 15–24 in the economically active population.
LUR	Long-term unemployment rate, defined as the percentage of unemployed persons who have been unemployed for 12 months or more.

```
data("EuroUnemployment", package = "mclust")
summary(EuroUnemployment)
##       TUR              YUR            LUR
## Min.   : 3.50   Min.   : 7.7   Min.   : 0.60
## 1st Qu.: 6.35   1st Qu.:15.4   1st Qu.: 2.10
## Median : 8.70   Median :20.5   Median : 3.70
## Mean   :10.05   Mean   :23.2   Mean   : 4.86
## 3rd Qu.:11.35   3rd Qu.:24.1   3rd Qu.: 6.80
## Max.   :26.50   Max.   :53.2   Max.   :19.50
```

The following code fits the isotropic EII and unconstrained VVV agglomerative hierarchical clustering models to this data:

```
HC_EII <- hc(EuroUnemployment, modelName = "EII")
HC_VVV <- hc(EuroUnemployment, modelName = "VVV")
```

Merge dendrograms (with uniform length between the hierarchy levels) for the EII and VVV models are plotted and shown in Figure 3.26.

```
plot(HC_EII, what = "merge", labels = TRUE, hang = 0.02)
plot(HC_VVV, what = "merge", labels = TRUE, hang = 0.02)
```

Log-likelihood dendrograms for the EII and VVV models are plotted and shown in Figure 3.27. These dendrograms extend from one cluster only as far as the classification likelihood increases and is defined (no singletons or singular covariances).

```
plot(HC_EII, what = "loglik")
plot(HC_VVV, what = "loglik")
```

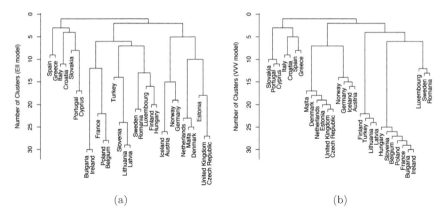

(a)   (b)

FIGURE 3.26: Dendrograms with height corresponding to the number of groups for model-based hierarchical clustering of the EuroUnemployment data with the EII (a) and VVV (b) models.

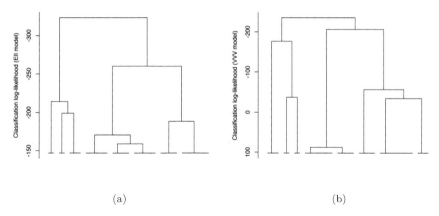

(a)   (b)

FIGURE 3.27: Dendrograms with height corresponding to the classification log-likelihood for model-based hierarchical clustering of the EuroUNemployment data with the EII (a) and VVV (b) models. The log-likelihood is undefined at the other levels of the trees due to the presence of singletons.

3.6.1   Agglomerative Clustering for Large Datasets

Model-based hierarchical agglomerative clustering is computationally demanding on large datasets, both in terms of time and memory. However, it is possible to initialize the modeling from a partition containing fewer groups than the number of observations. For example, Posse (2001) proposed using a subgraph of the minimum spanning tree associated with the data to derive an initial partition for model-based hierarchical clustering. He illustrated the method by applying it to a multiband MRI image of the human brain and to data on global precipitation climatology.

The optional argument partition available in the function hc() allows us to specify an initial partition of observations from which to start the agglomeration process.

EXAMPLE 3.10: Model-based hierarchical clustering on star data

As a simple illustration, consider the HRstars dataset from the **GDAdata** package (Unwin, 2015), which contains 6,200 measurements of star data on four variables. We first use k-means to divide the data into 100 groups, and then apply MBAHC to the data with this initial partition.

```
data("HRstars", package = "GDAdata")
set.seed(0)
initial <- kmeans(HRstars[, -1], centers = 100, nstart=10)$cluster
HC_VVV <- hc(HRstars[, -1], modelName = "VVV",
             partition = initial, use = "VARS")
HC_VVV
## Call:
## hc(data = HRstars[, -1], modelName = "VVV", use = "VARS", partition =
## initial)
##
## Model-Based Agglomerative Hierarchical Clustering
## Model name       = VVV
## Use              = VARS
## Number of objects = 6220
```

3.7   Initialization in mclust

As discussed in Section 2.2.3, initialization of the EM algorithm in model-based clustering is often crucial, because the likelihood surfaces for models of interest tend to have many local maxima and may even be unbounded, not to mention having regions where progress is slow. In **mclust**, the EM algorithm for multivariate data is initialized with the partitions obtained from model-based agglomerative hierarchical clustering (MBAHC), by default using the unconstrained (VVV) model. Efficient numerical algorithms for agglomerative hierarchical clustering with multivariate normal models have been discussed by Fraley (1998) and briefly reviewed in Section 3.6. In this approach, the two clusters are merged that yield the minimum regularized merge criterion (derived from the classification likelihood) over all possible merges at the current stage of the procedure.

Using unconstrained MBAHC is particularly convenient for GMMs because the underlying probability model (VVV) is shared to some extent by both the initialization step and the model fitting step. This strategy also has the advantage in that it provides the basis for initializing EM for any number of mixture components and component-covariance parameterizations. Although there is no guarantee that EM when so initialized will converge to a finite local optimum, it often provides reasonable starting values.

An issue with hierarchical clustering methods is that the resolution of ties in the merge criterion can have a significant effect on the downstream clustering outcome. Such ties are not uncommon when the data are discrete or when continuous data are rounded. Moreover, as the merges progress, these ties may not be exact for numerical reasons, and as a consequence results may be sensitive to relatively small changes in the data, including the order of the variables.

One way of detecting sensitivity in the results of a clustering method is to apply it to perturbed data. Such perturbations can be implemented in a number of ways, such as jittering the data, changing the order of the variables, or comparing results on different subsets of the data when there are a sufficient number of observations. Iterative methods, such as EM, can be initialized with different starting values.

EXAMPLE 3.11: Clustering flea beatle data

As an illustration of the issues related to the initialization of the EM algorithm, consider the flea beatle data available in package **tourr** (Wickham and Cook, 2022). This dataset provides six physical measurements for a sample of 74 fleas from three species as shown in the table below.

Variable	Description
species	Species of flea beetle from the genus Chaetocnema: concinna, heptapotamica, heikertingeri.
tars1	Width of the first joint of the first tarsus in microns (the sum of measurements for both tarsi)
tars2	Width of the second joint of the first tarsus in microns (the sum of measurements for both tarsi)
head	Maximal width of the head between the external edges of the eyes (in 0.01 mm).
ade1	Maximal width of the aedeagus in the fore-part (in microns).
ade2	Front angle of the aedeagus (1 unit = 7.5 degrees)
ade3	Aedeagus width from the side (in microns).

```
data("flea", package = "tourr")
X <- data.matrix(flea[, 1:6])
Class <- factor(flea$species,
              labels = c("Concinna", "Heikertingeri", "Heptapotamica"))
table(Class)
## Class
##      Concinna Heikertingeri Heptapotamica
##            21            31            22
```

Figure 3.28, obtained with the code below, shows the scatterplot matrix of
variables in the flea dataset with points marked according to the flea species.
Since the observed values are rounded (to the nearest integer presumably), there
is a considerable overplotting of points. For this reason, we added transparency
(or opacity) to cluster colors via the alpha channel in the following plot:

```
col <- mclust.options("classPlotColors")
clp <- clPairs(X, Class, lower.panel = NULL, gap = 0,
              symbols = c(16, 15, 17),
              colors = adjustcolor(col, alpha.f = 0.5))
clPairsLegend(x = 0.1, y = 0.3, class = clp$class,
             col = col, pch = clp$pch,
             title = "Flea beetle species")
```

Scrucca et al. (2016) discussed model-based clustering for this dataset and
showed that using the original variables for the MBAHC initialization step
leads to sub-optimal results in Mclust().

```
# set the default for the current session
mclust.options("hcUse" = "VARS")
mod1 <- Mclust(X)
# or specify the initialization method only for this model
# mod1 <- Mclust(X, initialization = list(hcPairs = hc(X, use = "VARS")))
summary(mod1)
## ----------------------------------------------------------
## Gaussian finite mixture model fitted by EM algorithm
## ----------------------------------------------------------
##
## Mclust EEE (ellipsoidal, equal volume, shape and orientation) model
## with 5 components:
##
##  log-likelihood  n df    BIC     ICL
##         -1292.3 74 55 -2821.3 -2825.8
##
## Clustering table:
##  1  2  3  4  5
```

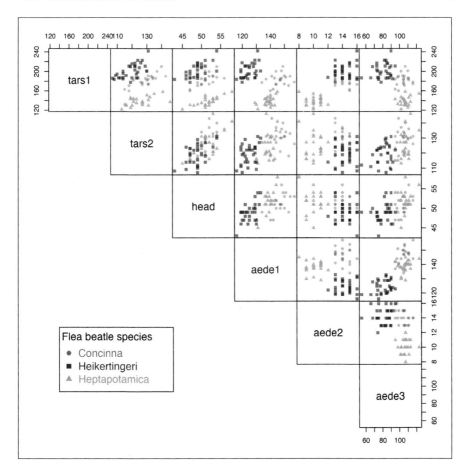

FIGURE 3.28: Scatterplot matrix for the flea dataset with points marked according to the true classes.

```
## 21   2 20 20 11
table(Class, mod1$classification)
##
## Class             1  2  3  4  5
##    Concinna      21  0  0  0  0
##    Heikertingeri  0  0  0 20 11
##    Heptapotamica  0  2 20  0  0
adjustedRandIndex(Class, mod1$classification)
## [1] 0.76757
```

If we reverse the order of the variables in the fit, MBAHC yields a different set of initial partitions for EM, a consequence of the discrete nature of the data.

With this initialization, Mclust still chooses the EEE model with 5 components, but with a different clustering partition.

```
mod2 <- Mclust(X[, 6:1])
summary(mod2)
## ----------------------------------------------------
## Gaussian finite mixture model fitted by FM algorithm
## ----------------------------------------------------
##
## Mclust EEE (ellipsoidal, equal volume, shape and orientation) model
## with 5 components:
##
##  log-likelihood  n df      BIC      ICL
##           -1287 74 55 -2810.8 -2812.7
##
## Clustering table:
##  1  2  3  4  5
## 22 21 22  7  2
table(Class, mod2$classification)
##
## Class             1  2  3  4  5
##    Concinna        0 21  0  0  0
##    Heikertingeri  22  0  0  7  2
##    Heptapotamica   0  0 22  0  0
adjustedRandIndex(Class, mod2$classification)
## [1] 0.81312
```

This second model has a higher BIC and better clustering accuracy.

In order to assess the stability of the results, we randomly start the EM algorithm using the function hcRandomPairs() to obtain a random hierarchical partitioning of the data for initialization:

```
mod3 <- Mclust(X, initialization = list(hcPairs = hcRandomPairs(X, seed = 1)))
summary(mod3)
## ----------------------------------------------------
## Gaussian finite mixture model fitted by EM algorithm
## ----------------------------------------------------
##
## Mclust VEE (ellipsoidal, equal shape and orientation) model with 4
## components:
##
##  log-likelihood  n df      BIC      ICL
##           -1297 74 51 -2813.5 -2814.3
##
## Clustering table:
```

```
## 1  2  3  4
## 27 22  4 21
table(Class, mod3$classification)
##
## Class           1  2  3  4
##   Concinna       0  0  0 21
##   Heikertingeri 27  0  4  0
##   Heptapotamica  0 22  0  0
adjustedRandIndex(Class, mod3$classification)
## [1] 0.90759
```

In this case, we obtain the VEE model with 4 components, which has a lower BIC but a higher ARI compared to the previous model. The function mclustBICupdate() can be used to retain the best models for each combination of number of components and cluster-covariance parameterization across several replications:

```
BIC <- NULL
for (i in 1:50)
{
  # get BIC table from initial random start
  BIC0 <- mclustBIC(X, verbose = FALSE,
                   initialization = list(hcPairs = hcRandomPairs(X)))
  # update BIC table by merging best BIC values for each
  # G and modelNames
  BIC  <- mclustBICupdate(BIC, BIC0)
}
summary(BIC, k = 5)
## Best BIC values:
##             EEE,3      VEE,3     EEE,4     VEE,4     EEE,5
## BIC      -2785.6 -2789.4245 -2796.966 -2805.838 -2811.541
## BIC diff     0.0    -3.8529   -11.394   -20.266   -25.969
```

We can then recover the best model object by a subsequent call to Mclust:

```
mod4 <- Mclust(X, x = BIC)
summary(mod4)
## ----------------------------------------------------
## Gaussian finite mixture model fitted by EM algorithm
## ----------------------------------------------------
##
## Mclust EEE (ellipsoidal, equal volume, shape and orientation) model
## with 3 components:
##
##  log-likelihood  n df     BIC     ICL
```

```
##             -1304.6 74 41 -2785.6 -2785.6
##
## Clustering table:
##  1  2  3
## 31 22 21
table(Class, mod4$classification)
##
## Class             1  2  3
##   Concinna        0  0 21
##   Heikertingeri  31  0  0
##   Heptapotamica   0 22  0
adjustedRandIndex(Class, mod4$classification)
## [1] 1
```

The above procedure is able to identify the true clustering structure, but it is a time-consuming process which for large datasets may not be feasible. Moreover, even when using multiple random starts, there is no guarantee that the best solution found is the best that can be achieved.

Scrucca and Raftery (2015) discussed several approaches to improve the hierarchical clustering initialization for model-based clustering. The main idea is to apply a transformation to the data in an effort to enhance separation among clusters before applying the MBAHC for the initialization step. The EM algorithm is then run using the data on the original scale. Among the studied transformations, one that often worked reasonably well (and is used as the default in **mclust**) is the *scaled SVD transformation*.

Let $\boldsymbol{X}$ be the $(n \times p)$ data matrix, and $\mathbb{X} = (\boldsymbol{X} - \mathbf{1}_n\bar{\boldsymbol{x}}^\top)\boldsymbol{S}^{-1/2}$ be the corresponding centered and scaled matrix, where $\bar{\boldsymbol{x}}$ is the vector of sample means, $\mathbf{1}_n$ is the unit vector of length n, and $\boldsymbol{S} = \mathrm{diag}(s_1^2, \ldots, s_d^2)$ is the diagonal matrix of sample variances. Consider the following singular value decomposition (SVD):

$$\mathbb{X} = \boldsymbol{U}\boldsymbol{\Omega}\boldsymbol{V}^\top = \sum_{i=1}^{r} \omega_i \boldsymbol{u}_i \boldsymbol{v}_i^\top,$$

where $\boldsymbol{u}_i$ are the left singular vectors, $\boldsymbol{v}_i$ the right singular vectors, $\omega_1 \geq \omega_2 \geq \cdots \geq \omega_r > 0$ the corresponding singular values, and $r \leq \min(n, d)$ the rank of matrix $\mathbb{X}$, with equality when there are no singularities. The scaled SVD transformation is computed as:

$$\boldsymbol{\mathcal{X}} = \mathbb{X}\boldsymbol{S}^{-1/2}\boldsymbol{V}\boldsymbol{\Omega}^{-1/2} = \boldsymbol{U}\boldsymbol{\Omega}^{1/2},$$

for which $\mathrm{E}(\boldsymbol{\mathcal{X}}) = \mathbf{0}$ and $\mathrm{Var}(\boldsymbol{\mathcal{X}}) = \boldsymbol{\Omega}/n = \mathrm{diag}(\omega_i)/n$. Thus, in the transformed scale the features are centered, uncorrelated, and with decreasing variances equal to the square root of the eigenvalues of the marginal sample correlation matrix.

EXAMPLE 3.12: Clustering flea beetle data (continued)

A GMM estimation initialized using MBAHC with the scaled SVD transformation of the data described above is obtained with the following code:

```
mclust.options("hcUse" = "SVD")  # restore the default
mod5 <- Mclust(X) # X is the unscaled flea data
# or specify only for this model fit
# mod5 <- Mclust(X, initialization = list(hcPairs = hc(X, use = "SVD")))
summary(mod5)
## --------------------------------------------------------
## Gaussian finite mixture model fitted by EM algorithm
## --------------------------------------------------------
##
## Mclust EEE (ellipsoidal, equal volume, shape and orientation) model
## with 3 components:
##
## log-likelihood n df     BIC     ICL
##         -1304.6 74 41 -2785.6 -2785.6
##
## Clustering table:
##  1  2  3
## 21 31 22
table(Class, mod5$classification)
##
## Class            1  2  3
##    Concinna      21  0  0
##    Heikertingeri  0 31  0
##    Heptapotamica  0  0 22
adjustedRandIndex(Class, mod5$classification)
## [1] 1
```

In this case, a single run with the scaled SVD initialization strategy (the default strategy for the Mclust() function) yields the highest BIC, and a perfect classification of the fleas into the actual species. However, as the true clustering would not usually be known, analysis with perturbations and/or multiple initialization strategies is always advisable. With EM, it is possible to do model-based clustering starting with parameter estimates, conditional probabilities, or classifications other than those produced by model-based agglomerative hierarchical clustering. The next section provides some further details on how to run an EM algorithm for Gaussian mixtures in **mclust**.

3.8   EM Algorithm in mclust

As described in Section 2.2.2, an iteration of EM consists of an E-step and an M-step. The E-step computes a matrix $\boldsymbol{Z} = \{z_{ik}\}$, where z_{ik} is an estimate of the conditional probability that observation i belongs to cluster k given the current parameter estimates. The M-step computes parameter estimates given $\boldsymbol{Z}$.

mclust provides functions em() and me() implementing the EM algorithm for maximum likelihood estimation in Gaussian mixture models for all 14 of the covariance parameterizations based on eigen-decomposition. The em() function starts with the E-step; besides the data and model specification, the model parameters (means, covariances, and mixing proportions) must be provided. The me() function, on the other hand, starts with the M-step; besides the data and model specification, the matrix of conditional probabilities must be provided. The output for both are the maximum likelihood estimates of the model parameters and the matrix of conditional probabilities.

Functions estep() and mstep() implement the individual E- and M- steps, respectively, of an EM iteration. Conditional probabilities and the log-likelihood can be recovered from parameters via estep(), while parameters can be recovered from conditional probabilities using mstep().

EXAMPLE 3.13: Single M- and E- steps using the iris data

Consider the well-known iris dataset (Anderson, 1935; Fisher, 1936) which contains the measurements (in cm) of sepal length and width, and petal length and width, for 50 flowers from each of 3 species of Iris, namely setosa, versicolor, and virginica.

The following code shows how single M- and E- steps can be performed using functions mstep() and estep():

```
data("iris", package = "datasets")
str(iris)
## 'data.frame': 150 obs. of  5 variables:
## $ Sepal.Length: num 5.1 4.9 4.7 4.6 5 5.4 4.6 5 4.4 4.9 ...
## $ Sepal.Width : num 3.5 3 3.2 3.1 3.6 3.9 3.4 3.4 2.9 3.1 ...
## $ Petal.Length: num 1.4 1.4 1.3 1.5 1.4 1.7 1.4 1.5 1.4 1.5 ...
## $ Petal.Width : num 0.2 0.2 0.2 0.2 0.2 0.4 0.3 0.2 0.2 0.1 ...
## $ Species : Factor w/ 3 levels "setosa","versicolor",..: 1 1 1 1 1 1 1
##     1 1 1 ...
ms <- mstep(iris[, 1:4], modelName = "VVV",
            z = unmap(iris$Species))
str(ms, 1)
## List of 7
## $ modelName : chr "VVV"
```

```
## $ prior : NULL
## $ n : int 150
## $ d : int 4
## $ G : int 3
## $ z : num [1:150, 1:3] 1 1 1 1 1 1 1 1 1 1 ...
## ..- attr(*, "dimnames")=List of 2
## $ parameters:List of 3
## - attr(*, "returnCode")= num 0
es <- estep(iris[, 1:4], modelName = "VVV",
            parameters = ms$parameters)
str(es, 1)
## List of 7
## $ modelName : chr "VVV"
## $ n : int 150
## $ d : int 4
## $ G : int 3
## $ z : num [1:150, 1:3] 1 1 1 1 1 1 1 1 1 1 ...
## ..- attr(*, "dimnames")=List of 2
## $ parameters:List of 3
## $ loglik : num -183
## - attr(*, "returnCode")= num 0
```

In this example, the initial estimate of z for the M-step is a matrix of indicator variables corresponding to a discrete classification taken from the true classes contained in iris$Species. Here, we used the function unmap() to convert the classification into the corresponding matrix of indicator variables. The inverse function, called map(), converts a matrix in which each row sums to 1 into an integer vector specifying for each row the column index of the maximum. This last operation is basically the MAP when applied to the matrix of posterior conditional probabilities (see Section 2.2.4).

3.9 Further Considerations in Cluster Analysis via Mixture Modeling

Clustering can be affected by control parameter settings, such as convergence tolerances within the clustering functions, although the defaults are often adequate.

Control parameters used by the EM functions in **mclust** are set and retrieved using the function emControl(). These include:

eps A tolerance value for terminating iterations due to ill-conditioning, such as near singularity in covariance matrices. By default this is set to

.Machine$double.eps, the relative machine precision, which has the value 2.220446e-16 on IEEE-compliant machines.

tol A vector of length 2 giving iteration convergence tolerances. By default this is set to c(1.0e-5, sqrt(.Machine$double.eps)). The first value is the tolerance for the relative convergence of the log-likelihood in the EM algorithm, and the second value is the relative convergence tolerance for the M-step of those models that have an iterative M-step, namely the models VEI, EVE, VEE, VVE, and VEV (see Table 2.2).

itmax An integer vector of length two giving integer limits on the number of EM iterations and on the number of iterations in the inner loop for models with iterative M-step (see above). By default this is set to c(.Machine$integer.max, .Machine$integer.max), allowing termination to be completely governed by the control parameter tol. A warning is issued if this limit is reached before the convergence criterion is satisfied.

Although these control settings are in a sense hidden by the defaults, they may have a significant effect on results in some instances and should be taken into consideration in any analysis.

The **mclust** implementation includes various warning messages in cases of potential problems. These are issued if mclust.options("warn") is set to TRUE or specified directly in function calls. Note, however, that by default mclust.options("warn") = FALSE, so that these warning messages are suppressed.

Finally, it is important to take into account numerical issues in model-based cluster analysis, and more generally in GMM estimation. The computations for estimating the model parameters break down when the covariance corresponding to one or more components becomes ill-conditioned (singular or nearly singular), and cannot proceed if clusters contain only a few observations or if the observations they contain are nearly colinear. Estimation may also fail when one or more mixing proportions shrink to negligible values. Including a prior is often helpful in such situations (see Section 7.2).

4

Mixture-Based Classification

Classification is an instance of supervised learning, where the class of each observation is known. Unlike in the unsupervised case, the main objective here is to build a classifier for classifying future observations. This chapter describes probabilistic classification following a mixture-based approach. It describes various Gaussian mixture models for supervised learning. The implementation available in **mclust** is presented using several data analysis examples. Different ways of assessing classifier performance are also discussed. The problem of unequal costs of misclassification and the classification with unbalanced classes is presented, followed by solutions implemented in **mclust**. The chapter concludes with an introduction to the semi-supervised classification problems, in which only some of the training data have known labels.

4.1 Classification as Supervised Learning

Chapter 3 discussed methods for clustering, an instance of *unsupervised learning*. There the main goal was to identify the presence of groups of homogeneous observations based on the measurements available for a set of variables or features. This chapter deals with the *supervised learning* problem, where the classification of each observation is known. In this case, the main objective is to build a classifier (or decision rule) for classifying future observations from the available data (Bishop, 2006; Hastie et al., 2009; Alpaydin, 2014). This task is also known by various other names, such as *statistical pattern recognition* or *discriminant analysis* (McLachlan, 2004).

In the probabilistic approach to classification, a statistical model is estimated to predict the class C_k for $k = 1, \ldots, K$ of a given observation with feature vector $\boldsymbol{x}$. This model provides a *posterior class probability* $\Pr(C_k|\boldsymbol{x})$ for each class, which can then be used to determine the class membership for new observations. Some modeling methods directly estimate posterior probabilities by constructing, or *learning*, a discriminant function $\eta_k(\boldsymbol{x}) = \Pr(C_k|\boldsymbol{x})$ that maps the features $\boldsymbol{x}$ directly onto a class C_k. These are called *discriminative models*, of which a popular instance is the logistic regression model for binary-class problems.

Other approaches try to explicitly or implicitly model the distribution of features as well as classes, and then obtain the posterior probabilities using Bayes' theorem. Thus, by learning the class-conditional densities $f(\boldsymbol{x}|C_k)$ and the prior class probabilities $\Pr(C_k)$ for each class C_k ($k = 1, \ldots, K$), the posterior class probabilities are given by

$$\Pr(C_k|\boldsymbol{x}) = \frac{f(\boldsymbol{x}|C_k)\Pr(C_k)}{\sum\limits_{g=1}^{K} f(\boldsymbol{x}|C_g)\Pr(C_g)} .$$

Methods that follow this approach, such as those based on finite mixture modeling, are called *generative models*.

Typically, classification models are estimated using the information from a *training set*, meaning a dataset used for learning or fitting the model in order to obtain parameter estimates. The same dataset, if used also for model tuning such as hyperparameter estimation or feature selection and for evaluating the classifier, tends to produce an optimistic assessment of performance. This phenomenon is called *overfitting*, meaning that there is a risk of fitting a model that too closely corresponds to a particular set of data, and therefore may fail to fit additional data or predict future observations well. For these reasons, it is advisable to perform model tuning using a *validation set*, a dataset designed for this purpose and usually set aside from the original dataset. An alternative is to repeatedly split into a training set and a validation set using resampling approaches, such as *cross-validation*, to be discussed in Section 4.4.3. Another subset of the original dataset is also set aside in advance as a *test set* for the final evaluation of the classifier.

4.2   Gaussian Mixture Models for Classification

Consider a training dataset $\mathcal{D}_{\text{train}} = \{(\boldsymbol{x}_1, y_1), \ldots, (\boldsymbol{x}_n, y_n)\}$ for which both the feature vectors $\boldsymbol{x}_i$ and the true classes $y_i \in \{C_1, \ldots, C_K\}$ are known. Each observation has an associated class label C_k.

Mixture-based classification models typically assume that the density within each class follows a Gaussian mixture distribution:

$$f(\boldsymbol{x}|C_k) = \sum_{g=1}^{G_k} \pi_{g,k} \phi(\boldsymbol{x}; \boldsymbol{\mu}_{g,k}, \boldsymbol{\Sigma}_{g,k}), \tag{4.1}$$

where G_k is the number of components within class k, $\pi_{g,k}$ are the mixing probabilities for class k ($\pi_{g,k} > 0$ and $\sum_{g=1}^{G_k} \pi_{g,k} = 1$), and $\boldsymbol{\mu}_{g,k}$ and $\boldsymbol{\Sigma}_{g,k}$ are, respectively, the mean vectors and the covariance matrices for component g within class k.

Hastie and Tibshirani (1996) proposed the *Mixture Discriminant Analysis*

(MDA) model where it is assumed that the covariance matrix is the same for all classes but is otherwise unconstrained ($\mathbf{\Sigma}_{gk} = \mathbf{\Sigma}$ for all $g = 1, \ldots, G_k$ and $k = 1, \ldots, K$ in equation (4.1)). Moreover, the number of mixture components is the same for each class and assumed known.

Bensmail and Celeux (1996) proposed the *Eigenvalue Decomposition Discriminant Analysis* (EDDA) model which assumes that the density for each class can be described by a single Gaussian component ($G_k = 1$ for all k in equation (4.1)), with the class covariance structure factorized as

$$\mathbf{\Sigma}_k = \lambda_k \mathbf{U}_k \mathbf{\Delta}_k \mathbf{U}_k^\top.$$

As for GMM clustering, several classification models can be obtained from the above decomposition. If each component has the same covariance matrix ($\mathbf{\Sigma}_k = \lambda \mathbf{U} \mathbf{\Delta} \mathbf{U}^\top$ — model EEE in Table 2.1), then EDDA is equivalent to the classical *Linear Discriminant Analysis* (LDA) model. If the component covariance matrices are unconstrained and vary between components ($\mathbf{\Sigma}_k = \lambda_k \mathbf{U}_k \mathbf{\Delta}_k \mathbf{U}_k^\top$ — model VVV in Table 2.1), then EDDA is equivalent to the *Quadratic Discriminant Analysis* (QDA) model. Finally, by assuming conditional independence of features within each class (models with coordinate axes orientation, denoted by ∗∗I in Table 2.1), the *Naïve-Bayes* models are obtained.

The most general model from equation (4.1) is the *MclustDA* model proposed by Fraley and Raftery (2002), which uses a finite mixture of Gaussian distributions within each class, in which the number of components and covariance matrix (parameterized by the eigen-decomposition described in Section 2.2.1) may differ among classes.

4.2.1 Prediction

Let τ_k be the class prior probability that an observation $\boldsymbol{x}$ comes from class C_k ($k = 1, \ldots, K$). By Bayes' theorem we can compute the posterior probability that an observation $\boldsymbol{x}$ belongs to class C_k as

$$\Pr(C_k|\boldsymbol{x}) = \frac{\tau_k f(\boldsymbol{x}|C_k)}{\sum_{j=1}^{K} \tau_j f(\boldsymbol{x}|C_j)}, \tag{4.2}$$

where $f(\boldsymbol{x}|C_k)$ is the probability density function in (4.1) specific to class C_k. As discussed earlier, this density depends on the assumed model for within-class distributions.

Thus an observation $\boldsymbol{x}$ can be classified according to the maximum a posteriori (MAP) rule to the class which has the highest posterior probability:

$$y = \{C_{\widehat{k}}\} \qquad \text{where} \quad \widehat{k} = \arg\max_k \Pr(C_k|\boldsymbol{x}) \propto \tau_k f(\boldsymbol{x}|C_k), \tag{4.3}$$

where the right-hand side follows by noting that the denominator in (4.2) is just a constant of normalization. This rule minimizes the expected misclassification rate and is known as the *Bayes classifier*.

4.2.2 Estimation

The parameters of the model in equation (4.1) can be estimated from the training dataset by maximum likelihood. In particular, for the EDDA model the parameters can be obtained with a single M-step from the EM algorithm for Gaussian mixtures described in Section 2.2.2, with z_{ik} set to 1 if observation i belongs to class k and 0 otherwise. For the general MclustDA model, as well as for MDA, a Gaussian mixture model can be estimated separately for each class using the EM algorithm, and parameters cannot be constrained across classes.

For the class prior probabilities, if the training data have been obtained by random sampling from the underlying population, the mixing proportions τ_k can be simply estimated by the sample proportions n_k/n, where n_k denotes the number of observations known to belong to class k and $n = \sum_{k=1}^{K} n_k$ is the number of observations in the training set. However, there are instances in which different values have to be assigned to the prior probabilities for the classes. This includes cases where the cost of misclassification may differ depending on the affected classes, and cases where classes have very different numbers of members. These issues are further discussed in Sections 4.6 and 4.5, respectively.

4.3 Classification in mclust

The main function available in **mclust** for classification tasks is MclustDA(), which requires a data frame or a matrix for the training data (data) and the corresponding vector of class labels (class). The type of mixture model to be fitted is specified by the argument modelType, a string that can take the values "MclustDA" (default) or "EDDA".

EXAMPLE 4.1: Classification of Wisconsin diagnostic breast cancer data

Consider a dataset of measurements for 569 patients on 30 features of the cell nuclei obtained from a digitized image of a fine needle aspirate (FNA) of a breast mass (Street et al., 1993; Mangasarian et al., 1995), available as one of several contributions to the 'Breast Cancer Wisconsin (Diagnostic) Data Set' of the UCI Machine Learning Repository (Dua and Graff, 2017). For each patient, the mass was diagnosed as either malignant or benign. This data can be obtained in **mclust** via the data command under the name wdbc. Following Mangasarian et al. (1995) and Fraley and Raftery (2002), we consider only three attributes in the following analysis: extreme area, extreme smoothness, and mean texture.

```
data("wdbc", package = "mclust")
X <- wdbc[, c("Texture_mean", "Area_extreme", "Smoothness_extreme")]
Class <- wdbc[, "Diagnosis"]
```

We randomly assign approximately two-thirds of the observations to the training set, and the remaining ones to the test set, as follows:

```
set.seed(123)
train <- sample(1:nrow(X), size = round(nrow(X)*2/3), replace = FALSE)
X_train <- X[train, ]
Class_train <- Class[train]
tab <- table(Class_train)
cbind(Counts = tab, "%" = prop.table(tab)*100)
##    Counts       %
## B     251 66.227
## M     128 33.773
X_test <- X[-train, ]
Class_test <- Class[-train]
tab <- table(Class_test)
cbind(Counts = tab, "%" = prop.table(tab)*100)
##    Counts       %
## B     106 55.789
## M      84 44.211
```

The distribution of the features with training data points marked according to cancer diagnosis is shown in Figure 4.1:

```
clp <- clPairs(X_train, Class_train, lower.panel = NULL)
clPairsLegend(0.1, 0.3, col = clp$col, pch = clp$pch,
              class = ifelse(clp$class == "B", "Benign", "Malign"),
              title = "Breast cancer diagnosis:")
```

The function MclustDA() provides fitting capabilities for the EDDA model by specifying the optional argument modelType = "EDDA". The corresponding function call is as follows:

```
mod1 <- MclustDA(X_train, Class_train, modelType = "EDDA")
summary(mod1)
## ------------------------------------------------
## Gaussian finite mixture model for classification
## ------------------------------------------------
##
## EDDA model summary:
##
## log-likelihood   n df      BIC
```

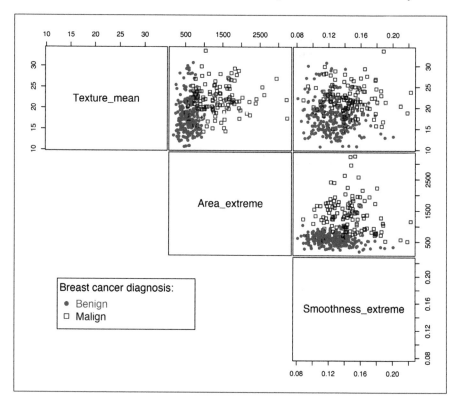

FIGURE 4.1: Pairwise scatterplot matrix of selected features for the breast cancer data with points distinguished by tumor diagnosis.

```
##             -2934.1 379 12 -5939.4
##
## Classes   n     % Model G
##       B 251 66.23    VVI 1
##       M 128 33.77    VVI 1
##
## Training confusion matrix:
##        Predicted
## Class   B   M
##     B 249   2
##     M  17 111
## Classification error = 0.0501
## Brier score         = 0.0374
```

The estimated EDDA mixture model with the largest BIC is the VVI model, in which each group is described by a single Gaussian component with varying volume and shape, and orientation aligned with the coordinate axes.

As mentioned earlier, this model is a member of the Naïve-Bayes family. By default, the summary() function also returns the confusion matrix obtained by cross-tabulation of the input and predicted classes, followed by two measures of accuracy to be discussed in Section 4.4.

Estimated parameters can be shown with the summary() function by setting the optional argument parameters as follows:

```
summary(mod1, parameters = TRUE)
## ------------------------------------------------
## Gaussian finite mixture model for classification
## ------------------------------------------------
##
## EDDA model summary:
##
##  log-likelihood   n df      BIC
##          -2934.1 379 12 -5939.4
##
## Classes    n     % Model G
##        B 251 66.23   VVI 1
##        M 128 33.77   VVI 1
##
## Class prior probabilities:
##       B       M
## 0.66227 0.33773
##
## Class = B
##
## Means:
##                          [,1]
## Texture_mean          17.95530
## Area_extreme         562.71673
## Smoothness_extreme     0.12486
##
## Variances:
## [,,1]
##                    Texture_mean Area_extreme Smoothness_extreme
## Texture_mean             15.312            0         0.00000000
## Area_extreme              0.000        26588         0.00000000
## Smoothness_extreme        0.000            0         0.00040151
##
## Class = M
##
## Means:
##                          [,1]
## Texture_mean          21.80203
```

```
## Area_extreme          1343.71094
## Smoothness_extreme       0.14478
##
## Variances:
## [,,1]
##                     Texture_mean Area_extreme Smoothness_extreme
## Texture_mean              12.408            0         0.00000000
## Area_extreme               0.000       288727         0.00000000
## Smoothness_extreme         0.000            0         0.00060343
##
## Training confusion matrix:
##        Predicted
## Class   B    M
##     B 249    2
##     M  17  111
## Classification error = 0.0501
## Brier score          = 0.0374
```

The confusion matrix and evaluation metrics for a new test set can be obtained by providing the data matrix of features (newdata) and the corresponding classes (newclass):

```
summary(mod1, newdata = X_test, newclass = Class_test)
## ------------------------------------------------
## Gaussian finite mixture model for classification
## ------------------------------------------------
##
## EDDA model summary:
##
##   log-likelihood   n df      BIC
##          -2934.1 379 12  -5939.4
##
## Classes   n      % Model G
##       B 251  66.23   VVI 1
##       M 128  33.77   VVI 1
##
## Training confusion matrix:
##        Predicted
## Class   B    M
##     B 249    2
##     M  17  111
## Classification error = 0.0501
## Brier score          = 0.0374
##
## Test confusion matrix:
```

```
##          Predicted
## Class   B   M
##     B 103   3
##     M   5  79
## Classification error = 0.0421
## Brier score          = 0.0357
```

Note that, for this model, the performance metrics on the test set are no worse than those for the training set, and in fact are even slightly better. This indicates that the estimated model is not overfitting the data, which is likely due to the parsimonious covariance model adopted.

Objects returned by MclustDA() can be visualized in a variety of ways through the associated plot method. For instance, the pairwise scatterplot matrix between the features, showing both the known classes and the estimated mixture components, is displayed in Figure 4.2 and obtained with the code:

```
plot(mod1, what = "scatterplot")
```

Figure 4.3 displays the pairwise scatterplots showing the misclassified training data points obtained with the following code:

```
plot(mod1, what = "error")
```

EDDA imposes a single mixture component for each group. However, in certain circumstances, a more flexible model may result in a better classifier. As mentioned in Section 4.2, a more general approach, called *MclustDA*, is available, in which a finite mixture of Gaussian distributions is used within each class, with both the number of components and covariance matrix structures (expressed following the usual eigen-decomposition in (2.4)) allowed to differ among classes. This is the model estimated by default (or by setting modelType = "MclustDA"):

```
mod2 <- MclustDA(X_train, Class_train)
summary(mod2, newdata = X_test, newclass = Class_test)
## ------------------------------------------------
## Gaussian finite mixture model for classification
## ------------------------------------------------
##
## MclustDA model summary:
##
##  log-likelihood   n df     BIC
##         -2893.6 379 26 -5941.6
##
## Classes   n      % Model G
##       B 251 66.23   EEI 3
```

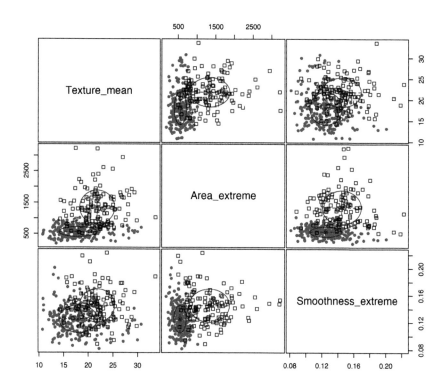

FIGURE 4.2: Pairwise scatterplot matrix of selected features for the breast cancer training data with points distinguished by observed classes and ellipses representing the Gaussian distribution estimated for each class by EDDA.

```
##        M 128 33.77    EVI 2
##
## Training confusion matrix:
##       Predicted
## Class   B    M
##     B  248   3
##     M    8  120
## Classification error = 0.029
## Brier score          = 0.0262
##
## Test confusion matrix:
##       Predicted
## Class   B    M
##     B  103   3
##     M    5   79
```

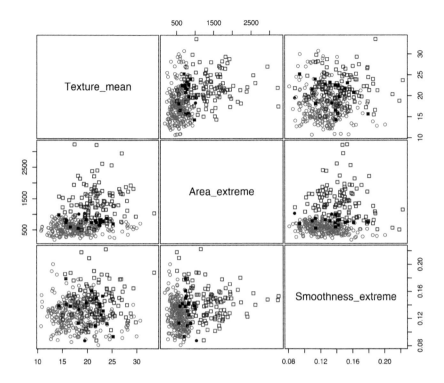

FIGURE 4.3: Pairwise scatterplot matrix of selected features for the breast cancer training data with points distinguished by observed classes and filled black points representing those cases misclassified by the fitted EDDA model.

```
## Classification error = 0.0421
## Brier score        = 0.0273
```

MclustDA fits a three-component EEI mixture to benign cases, and a two-component EVI mixture to the malignant cases. Note that diagonal covariance structures are used within each class. The training classification error rate is smaller for this model than for the EDDA model. However, the test mis-classification rate is the same for the two types of models. This is an effect of *overfitting* induced by the increased complexity of the MclustDA model, which has 26 parameters to estimate, more than twice the number required by the EDDA model.

Figure 4.4 displays a matrix of pairwise scatterplots between the features showing both the known classes and the estimated mixture components drawn with the code:

```
plot(mod2, what = "scatterplot")
```

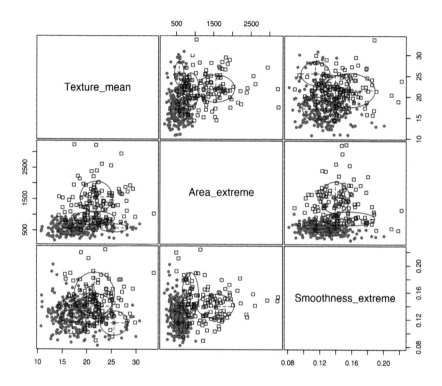

FIGURE 4.4: Scatterplots of selected features for the breast cancer training data with points distinguished by observed classes and ellipses representing the Gaussian distribution estimated for each class by MclustDA.

Specific marginals can be obtained by using the optional argument dimens. For instance, the following code produces the scatterplot for the first two features:

```
plot(mod2, what = "scatterplot", dimens = c(1, 2))
```

Finally, note that the MDA model of Hastie and Tibshirani (1996) is equivalent to MclustDA with $\Sigma_k = \lambda U \Delta U^\top$ (model EEE) and fixed $G_k \geq 1$ for each class. For instance, an MDA model with two mixture components for each class can be fitted using the code:

```
mod3 <- MclustDA(X_train, Class_train, G = 2, modelNames = "EEE")
summary(mod3, newdata = X_test, newclass = Class_test)
```

```
## -------------------------------------------------
## Gaussian finite mixture model for classification
## -------------------------------------------------
##
## MclustDA model summary:
##
##  log-likelihood   n df   BIC
##         -2910.8 379 26 -5976
##
## Classes   n     % Model G
##        B 251 66.23   EEE 2
##        M 128 33.77   EEE 2
##
## Training confusion matrix:
##         Predicted
## Class   B   M
##     B 247   4
##     M  10 118
## Classification error = 0.0369
## Brier score          = 0.0297
##
## Test confusion matrix:
##         Predicted
## Class   B   M
##     B 104   2
##     M   4  80
## Classification error = 0.0316
## Brier score          = 0.0248
```

4.4 Evaluating Classifier Performance

Evaluating the performance of a classifier is an essential part of any classification task. Mixture models used for classification are able to generate two types of predictions: the most likely class label according to the MAP rule in equation (4.3), and the posterior probabilities of class membership from equation (4.2). **mclust** automatically computes one performance measure for each type of prediction, namely the *misclassification error* and the *Brier score*, to be described in the following sections. Note, however, that several other measures exist, such as the Receiving Operating Characteristic (ROC) curve and the Area Under the [ROC] Curve (AUC) for two-class cases, and

it is possible to compute them using the estimates provided by the predict()
method associated with MclustDA objects.

4.4.1 Evaluating Predicted Classes: Classification Error

The simplest measure available is the *misclassification error rate*, or simply
the *classification error*, which is the proportion of wrong predictions made by
the classifier:

$$\text{CE} = \frac{1}{n} \sum_{i=1}^{n} I(\widehat{y}_i \neq y_i), \tag{4.4}$$

where $y_i = \{C_k\}$ is the known class for the ith observation, $\widehat{y}_i = \{C_{\widehat{k}}\}$ is the
predicted class label, and $I(\widehat{y}_i \neq y_i)$ is an indicator function that equals 1
if $\widehat{y}_i \neq y_i$ and 0 otherwise. A good classifier should have a small error rate,
preferably close to zero. Equivalently, the *accuracy* of a classifier is defined
as the proportion of correct predictions, $1 - \text{CE}$. Note, however, that when
classes are *unbalanced* (not represented more or less equally), the error rate
or accuracy may not be meaningful. A classifier could have a high overall
accuracy yet not be able to accurately detect members of small classes.

4.4.2 Evaluating Class Probabilities: Brier Score

The Brier score is a measure of the predictive accuracy for probabilistic
predictions. It is computed as the mean squared difference between the true
class indicators and the predicted probabilities.

Based on the original multi-class definition by Brier (1950), the following
formula provides the normalized Brier score:

$$\text{BS} = \frac{1}{2n} \sum_{i=1}^{n} \sum_{k=1}^{K} (C_{ik} - \widehat{p}_{ik})^2,$$

where n is the number of observations, K is the number of classes, $C_{ik} = 1$
if observation i is from class k and 0 otherwise, and $\widehat{p}_{ik}$ is the predicted
probability that observation i belongs to class k. In this formula, the inclusion
of the constant 2 in the denominator ensures that the index takes values in
the range $[0, 1]$ (Kruppa et al., 2014a,b).

The Brier score is a strictly proper score (Gneiting and Raftery, 2007), which
implies that it takes its minimal value only when the predicted probabilities
match the empirical probabilities. Thus, small values of the Brier score indicate
high prediction accuracy, with $\text{BS} = 0$ when the observations are all correctly
classified with probability one.

4.4.3 Estimating Classifier Performance: Test Set and Resampling-Based Validation

Any performance measure computed using the training set will tend to provide an optimistic performance estimate. For instance, the training classification error rate CE_{train} is obtained by applying equation (4.4) to the training observations. This measure of the accuracy of a classifier is optimistic because the same set of observations is used for both model estimation and for its assessment. A more realistic estimate can be obtained by computing the test misclassification error rate, which is the error rate computed on a fresh test set of m observations $\mathcal{D}_{\text{test}} = \{(\boldsymbol{x}_1^*, y_1^*), \ldots, (\boldsymbol{x}_m^*, y_m^*)\}$:

$$\text{CE}_{\text{test}} = \frac{1}{m} \sum_{i=1}^{m} I(y_i^* \neq \widehat{y}_i^*),$$

where $\widehat{y}_i^*$ is the predicted class label that results from applying the classifier with feature vector $\boldsymbol{x}^*$. This seems an obvious choice, "but, to get reasonable precision of the performance values, the size of the test set may need to be large" (Kuhn and Johnson, 2013, p. 66). The same considerations also apply to the Brier score.

In cases where a test set is not available, or its size is not sufficient to guarantee reliable estimates, an assessment of a model's performance can be obtained by resampling methods. Different resampling schemes are available, but all rely on modeling repeated samples drawn from a training set.

Cross-validation is a simple and intuitive way to obtain a realistic performance measure. A standard resampling scheme is the V-fold cross-validation approach, which randomly splits the set of training observations into V parts or *folds*. At each step of the procedure, data from $V - 1$ folds are used for model fitting, and the held-out fold is used as a validation set. Figure 4.5 provides a schematic view of 10-fold cross-validation.

Consider a generic loss function of the prediction error, say $L(y, \widehat{y})$, that we would like to minimize. For instance, by setting $L(y, \widehat{y}) = I(y \neq \widehat{y})$, the $0-1$ loss, we obtain the misclassification error rate, whereas by setting $L(y, \widehat{y}) = (C_k - \widehat{p})^2$, the squared error with respect to the estimated probability $\widehat{p}$, we obtain the Brier score. The V-fold cross-validation steps are the following:

1. Split the training set into V folds of roughly equal size; (and stratified[1]), say $F_1, \ldots, F_V$.

2. For $v = 1, \ldots, V$:

   (a) fit the model using $\{(\boldsymbol{x}_i, y_i) : i \notin F_v\}$ as training set;

---

[1]In stratified cross-validation the data are randomly split in such a way that maintains the same class distribution in each fold. This is of particular relevance in the case of unbalanced class distribution. Furthermore, it has been noted that "stratification is generally a better scheme, both in terms of bias and variance, when compared to regular cross-validation" (Kohavi, 1995).

(b) evaluate the model using $\{(\boldsymbol{x}_i, y_i) : i \in F_v\}$ as validating set by computing

$$L_v = \frac{1}{n_v} \sum_{i \in F_v} L(y_i, \widehat{y}_i),$$

where n_v is the number of observations in fold F_v'.

3. Average the loss function over the folds by computing

$$L_{\mathrm{CV}} = \sum_{v=1}^{V} \frac{n_v}{n} L_v = \frac{1}{n} \sum_{v=1}^{V} \sum_{i \in F_v} L(y_i, \widehat{y}_i).$$

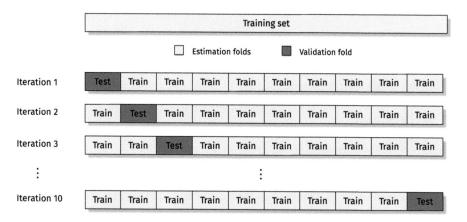

FIGURE 4.5: Schematic representation of the 10-fold cross-validation resampling technique.

There is no general rule for choosing an optimal value for V. If $V = n$, the procedure is called *leave-one-out cross-validation* (LOOCV), because one data point is held out at a time. Large values of V reduce the bias of the estimator but increase its variance, while small values of V increase the bias but decrease the variance. Furthermore, for large values of V, the computational burden may be quite high. For these reasons, it is often suggested to set V equal to 5 or 10 (Hastie et al., 2009, Sec. 7.10).

An advantage of using a cross-validation approach is that it provides an estimate of the standard error of the procedure. This can be computed as

$$\mathrm{se}(L_{\mathrm{CV}}) = \frac{\mathrm{sd}(L)}{\sqrt{V}},$$

where

$$\mathrm{sd}(L) = \sqrt{\frac{\sum_{v=1}^{V} (L_v - L_{\mathrm{CV}})^2 n_v}{n(V-1)/V}}.$$

The estimate se(L_{CV}) is often used for implementing the *one-standard error rule*: when models of different complexity are compared, select the simplest model whose performance is within one standard error of the best value (Breiman et al., 1984, Sections 8.1 and 14.1). For an in-depth investigation of the behavior of cross-validation for some commonly used statistical models, see Bates et al. (2021).

4.4.4 Cross-Validation in mclust

The function `cvMclustDA()` is available in **mclust** to carry out *V*-fold cross-validation as discussed above. It requires an object as returned by `MclustDA()` and, among the optional arguments, `nfold` can be used to set the number of folds (by default set to 10).

EXAMPLE 4.2: Evaluation of classification models using cross-validation for the Wisconsin diagnostic breast cancer data

Consider the classification models estimated in Example 4.1. The following code computes the 10-fold CV for the selected EDDA and MclustDA models:

```
cv1 <- cvMclustDA(mod1)
str(cv1)
## List of 6
## $ classification: Factor w/ 2 levels "B","M": 2 1 1 1 1 2 1 1 1 1 ...
## $ z : num [1:379, 1:2] 0.191 0.744 0.972 0.927 0.705 ...
## ..- attr(*, "dimnames")=List of 2
## .. ..$ : NULL
## .. ..$ : chr [1:2] "B" "M"
## $ ce : num 0.0528
## $ se.ce : num 0.0125
## $ brier : num 0.0399
## $ se.brier : num 0.00734
cv2 <- cvMclustDA(mod2)
str(cv2)
## List of 6
## $ classification: Factor w/ 2 levels "B","M": 2 1 1 1 2 2 1 1 1 2 ...
## $ z : num [1:379, 1:2] 0.304 0.891 0.916 0.994 0.471 ...
## ..- attr(*, "dimnames")=List of 2
## .. ..$ : NULL
## .. ..$ : chr [1:2] "B" "M"
## $ ce : num 0.0317
## $ se.ce : num 0.00765
## $ brier : num 0.0298
## $ se.brier : num 0.00462
```

The list of values returned by `cvMclustDA()` contains the cross-validated predicted classes (classification), the posterior class conditional probabilities (z), followed by the cross-validated metrics (the misclassification error rate and the Brier score) and their standard errors. The latter can be extracted using:

```
unlist(cv1[c("ce", "se.ce", "brier", "se.brier")])
##        ce      se.ce     brier  se.brier
## 0.0527704 0.0124978 0.0399285 0.0073402
unlist(cv2[c("ce", "se.ce", "brier", "se.brier")])
##        ce      se.ce     brier  se.brier
## 0.0316623 0.0076524 0.0297516 0.0046195
```

Training and resampling metrics for all of the EDDA models and the MclustDA model can be obtained using the following code:

```
models <- mclust.options("emModelNames")
tab_CE <- tab_Brier <-
  matrix(as.double(NA), nrow = length(models)+1, ncol = 5)
rownames(tab_CE) <- rownames(tab_Brier) <-
  c(paste0("EDDA[", models, "]"), "MCLUSTDA")
colnames(tab_CE) <- colnames(tab_Brier) <-
  c("Train", "10-fold CV", "se(CV)", "lower", "upper")
for (i in seq(models))
{
  mod <- MclustDA(X, Class, modelType = "EDDA",
                  modelNames = models[i], verbose = FALSE)
  pred <- predict(mod, X)
  cv <- cvMclustDA(mod, nfold = 10, verbose = FALSE)
  #
  tab_CE[i, 1] <- classError(pred$classification, Class)$errorRate
  tab_CE[i, 2] <- cv$ce
  tab_CE[i, 3] <- cv$se.ce
  tab_CE[i, 4] <- cv$ce - cv$se.ce
  tab_CE[i, 5] <- cv$ce + cv$se.ce
  #
  tab_Brier[i, 1] <- BrierScore(pred$z, Class)
  tab_Brier[i, 2] <- cv$brier
  tab_Brier[i, 3] <- cv$se.brier
  tab_Brier[i, 4] <- cv$brier - cv$se.brier
  tab_Brier[i, 5] <- cv$brier + cv$se.brier
}
i <- length(models)+1
mod <- MclustDA(X, Class, modelType = "MclustDA", verbose = FALSE)
pred <- predict(mod, X)
cv <- cvMclustDA(mod, nfold = 10, verbose = FALSE)
#
```

```
tab_CE[i, 1] <- classError(pred$classification, Class)$errorRate
tab_CE[i, 2] <- cv$ce
tab_CE[i, 3] <- cv$se.ce
tab_CE[i, 4] <- cv$ce - cv$se.ce
tab_CE[i, 5] <- cv$ce + cv$se.ce
#
tab_Brier[i, 1] <- BrierScore(pred$z, Class)
tab_Brier[i, 2] <- cv$brier
tab_Brier[i, 3] <- cv$se.brier
tab_Brier[i, 4] <- cv$brier - cv$se.brier
tab_Brier[i, 5] <- cv$brier + cv$se.brier
```

The following table gives the training error, the 10-fold CV error with its standard error, and the lower and upper bounds computed as ± one standard error from the CV estimate:

```
tab_CE
##              Train 10-fold CV    se(CV)    lower    upper
## EDDA[EII] 0.112478   0.117750 0.0168034 0.100947 0.134554
## EDDA[VII] 0.079086   0.079086 0.0120644 0.067022 0.091150
## EDDA[EEI] 0.087873   0.087873 0.0089230 0.078950 0.096797
## EDDA[VEI] 0.091388   0.093146 0.0128794 0.080266 0.106025
## EDDA[EVI] 0.066784   0.072056 0.0083953 0.063661 0.080452
## EDDA[VVI] 0.043937   0.047452 0.0064740 0.040978 0.053926
## EDDA[EEE] 0.066784   0.068541 0.0076137 0.060928 0.076155
## EDDA[VEE] 0.072056   0.073814 0.0159524 0.057861 0.089766
## EDDA[EVE] 0.059754   0.065026 0.0073839 0.057642 0.072410
## EDDA[VVE] 0.042179   0.043937 0.0088144 0.035122 0.052751
## EDDA[EEV] 0.047452   0.050967 0.0075542 0.043412 0.058521
## EDDA[VEV] 0.052724   0.056239 0.0108316 0.045407 0.067071
## EDDA[EVV] 0.045694   0.047452 0.0074344 0.040017 0.054886
## EDDA[VVV] 0.036907   0.038664 0.0090412 0.029623 0.047705
## MCLUSTDA  0.022847   0.036907 0.0105586 0.026348 0.047465
```

The same information is also shown graphically in Figure 4.6 using:

```
library("ggplot2")
df <- data.frame(rownames(tab_CE), tab_CE)
colnames(df) <- c("model", "train", "cv", "se", "lower", "upper")
df$model <- factor(df$model, levels = rev(df$model))
ggplot(df, aes(x = model, y = cv, ymin = lower, ymax = upper)) +
  geom_point(aes(shape = "s1", color = "c1")) +
  geom_errorbar(width = 0.5, col = "dodgerblue3") +
  geom_point(aes(y = train, shape = "s2", color = "c2")) +
  scale_y_continuous(breaks = seq(0, 0.2, by = 0.01), lim = c(0, NA)) +
```

```
scale_color_manual(name = "",
                   breaks = c("c1", "c2"),
                   values = c("dodgerblue3", "black"),
                   labels = c("CV", "Train")) +
scale_shape_manual(name = "",
                   breaks = c("s1", "s2"),
                   values = c(19, 0),
                   labels = c("CV", "Train")) +
ylab("Classification error") + xlab("") + coord_flip() +
theme(legend.position = "top")
```

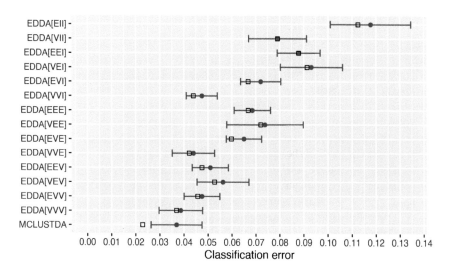

FIGURE 4.6: Training and cross-validated misclassification error rates of Gaussian mixture classification models for the breast cancer data.

The analogous table for the Brier score is:

```
tab_Brier
##             Train 10-fold CV    se(CV)    lower     upper
## EDDA[EII] 0.095691   0.096282 0.0141399 0.082142 0.110422
## EDDA[VII] 0.072572   0.072936 0.0099342 0.063001 0.082870
## EDDA[EEI] 0.055086   0.055573 0.0055047 0.050068 0.061078
## EDDA[VEI] 0.060386   0.062123 0.0094198 0.052703 0.071543
## EDDA[EVI] 0.047277   0.049384 0.0067815 0.042603 0.056166
## EDDA[VVI] 0.036197   0.037103 0.0044475 0.032655 0.041550
```

```
## EDDA[EEE] 0.051470    0.052056 0.0040711 0.047984 0.056127
## EDDA[VEE] 0.052813    0.054901 0.0085825 0.046318 0.063483
## EDDA[EVE] 0.043315    0.046130 0.0053728 0.040757 0.051502
## EDDA[VVE] 0.035243    0.035845 0.0065776 0.029268 0.042423
## EDDA[EEV] 0.039127    0.040170 0.0040287 0.036141 0.044199
## EDDA[VEV] 0.041625    0.043191 0.0074828 0.035708 0.050674
## EDDA[EVV] 0.036139    0.037555 0.0050249 0.032530 0.042580
## EDDA[VVV] 0.028803    0.030183 0.0063246 0.023859 0.036508
## MCLUSTDA  0.022999    0.026530 0.0064555 0.020075 0.032986
```

with the corresponding plot in Figure 4.7:

```
df <- data.frame(rownames(tab_Brier), tab_Brier)
colnames(df) <- c("model", "train", "cv", "se", "lower", "upper")
df$model <- factor(df$model, levels = rev(df$model))
ggplot(df, aes(x = model, y = cv, ymin = lower, ymax = upper)) +
  geom_point(aes(shape = "s1", color = "c1")) +
  geom_errorbar(width = 0.5, col = "dodgerblue3") +
  geom_point(aes(y = train, shape = "s2", color = "c2")) +
  scale_y_continuous(breaks = seq(0, 0.2, by = 0.01), lim = c(0, NA)) +
  scale_color_manual(name = "",
                     breaks = c("c1", "c2"),
                     values = c("dodgerblue3", "black"),
                     labels = c("CV", "Train")) +
  scale_shape_manual(name = "",
                     breaks = c("s1", "s2"),
                     values = c(19, 0),
                     labels = c("CV", "Train")) +
  ylab("Brier score") + xlab("") + coord_flip() +
  theme(legend.position = "top")
```

The plots in Figures 4.6 and 4.7 show that, by increasing the complexity of the model, it is sometimes possible to improve the accuracy of the predictions. In particular, the MclustDA model had better performance than the EDDA models, with the exception of the EDDA model with unconstrained covariances (VVV), which had both misclassification error rate and Brier score within one standard error from the best.

The information returned by cvMclustDA() can also be used for computing other cross-validation metrics. For instance, in the binary class case two popular measures are *sensitivity* and *specificity*. The sensitivity, or true positive rate, is given by the ratio of the number observations that are classified in the positive class to the total number of positive cases. The specificity, or true negative rate, is given by the ratio of observations that are classified in the negative class to the total number of negative cases. Both metrics are easily computed

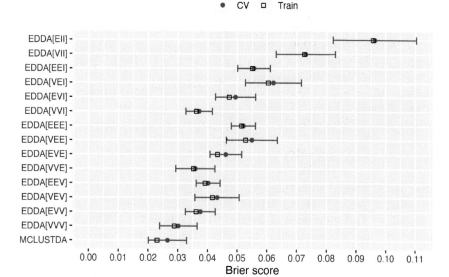

FIGURE 4.7: Training and cross-validated Brier scores of Gaussian mixture classification models for the breast cancer data.

from the *confusion matrix*, obtained by cross-tabulating the true classes and the classes predicted by a classifier.

EXAMPLE 4.3: ROC-AUC analysis of classification models for the Wisconsin diagnostic breast cancer data

Returning to the Wisconsin diagnostic breast cancer data in Example 4.1, we are mainly interested in the identification of patients with malignant diagnosis, so the positive class can be taken to be class M, while benign cases (B) are assigned to the negative class.

```
# confusion matrix
(tab <- table(Predict = cv1$classification, Class = Class_train))
##        Class
## Predict  B    M
##      B 248   17
##      M   3  111
tab[2, 2]/sum(tab[, 2])  # sensitivity
## [1] 0.86719
tab[1, 1]/sum(tab[, 1])  # specificity
## [1] 0.98805
```

The code above uses the cross-validated classifications obtained using the

MAP approach, which for the binary class case is equivalent to setting the classification probability threshold at 0.5. However, the posterior probabilities returned by cvMclustDA() can be used to get classifications at different threshold values. The following code computes the sensitivity and specificity over a fine grid of threshold values:

```
threshold <- seq(0, 1, by = 0.01)
sensitivity <- specificity <- rep(NA, length(threshold))
for(i in 1:length(threshold))
{
  pred <- factor(ifelse(cv1$z[, "M"] > threshold[i], "M", "B"),
                levels = c("B", "M"))
  tab <- table(pred, Class_train)
  sensitivity[i] <- tab[2, 2]/sum(tab[, 2])
  specificity[i] <- tab[1, 1]/sum(tab[, 1])
}
```

The metrics computed above for varying thresholds in binary decisions can be represented graphically using the *Receiver Operating Characteristic* (ROC) curve, which plots the sensitivity (true positive rate) vs. one minus the specificity (false positive rate). The resulting display is shown in Figure 4.8a and obtained with the following code:

```
plot(1-specificity, sensitivity, type = "l", lwd = 2)  # ROC curve
abline(h = c(0, 1), v = c(0, 1), lty = 3)  # limits of [0,1]x[0,1] region
abline(a = 0, b = 1, lty = 2)  # line of random classification
```

Notice that the optimal ROC curve would pass through the upper left corner, which corresponds to both sensitivity and specificity equal to 1.

A summary of the ROC curve which is used for evaluating the overall performance of a classifier is the *Area Under the Curve* (AUC). This is equal to 1 for a perfect classifier, and 0.5 for a random classification; values larger than 0.8 are considered to be good (Lantz, 2019, p. 333). Provided that the threshold grid is fine enough, a simple approximation of the AUC is obtained using the following function:

```
auc_approx <- function(tpr, fpr)
{
  x <- 1 - fpr
  y <- tpr
  dx <- c(diff(x), 0)
  dy <- c(diff(y), 0)
  sum(y * dx) + sum(dy * dx)/2
}
```

```
auc_approx(tpr = sensitivity, fpr = 1 - specificity)
## [1] 0.98129
```

The value of AUC indicates a classifier with a very good classification performance.

The same ROC-AUC analysis can also be replicated for the selected MclustDA model (in object mod2), producing the ROC curve shown in Figure 4.8b. The corresponding value of the AUC is 0.98506.

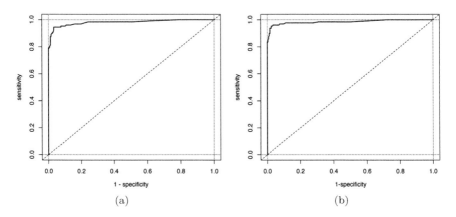

(a)   (b)

FIGURE 4.8: ROC curves from the cross-validated predictions of selected (a) EDDA and (b) MclustDA models for the breast cancer data.

Finally, the ROC curve can also be used to select an optimal threshold value. This can be set at the value where the true positive rate is high and the false positive rate is low, which is equivalent to maximizing *Youden's index*:

$$J = \text{Sensitivity} - (1 - \text{Specificity}) = \text{Sensitivity} + \text{Specificity} - 1.$$

Note that J corresponds to the vertical distance between the ROC curve and the random classification line. The following code computes Youden's index and the optimal threshold for the selected EDDA model:

```
J <- sensitivity + specificity - 1
threshold[which.max(J)]      # optimal threshold
## [1] 0.28
sensitivity[which.max(J)]    # sensitivity at optimal threshold
## [1] 0.94531
specificity[which.max(J)]    # specificity at optimal threshold
## [1] 0.96813
```

ROC curves are a suitable measure of performance when the distribution of the two classes is approximately equal. Otherwise, Precision-Recall (PR) curves are a better alternative. Both measures require a sufficient number of thresholds to obtain an accurate estimate of the corresponding area under the curve. A discussion of the relationship between ROC and PR curves can be found in Davis and Goadrich (2006). R implementations include CRAN package **PRROC** (Keilwagen et al., 2014; Grau et al., 2015) and **ROCR** (Sing et al., 2005)

4.5 Classification with Unequal Costs of Misclassification

In many practical applications, different costs are associated with different types of classification error. Thus it can be argued that the decision rule should be based on the principle that the total cost of misclassification should be minimized. Costs can be incorporated into a decision rule either at the learning stage or at the prediction stage.

We now describe a strategy for Gaussian mixtures that takes costs into account only at the final prediction stage. Let $c(k|j)$ be the cost of allocating an observation from class j to class $k \neq j$, with $c(j|j) = 0$. Let $p(k|j) = \Pr(C_k | \boldsymbol{x} \in C_j)$ be the probability of allocating an observation coming from class j to class k. The expected cost of misclassification (ECM) for class j is given by

$$\text{ECM}(j) = \sum_{k \neq j}^{K} c(k|j) p(k|j) \propto \sum_{k \neq j}^{K} c(k|j) \tau_k f(\boldsymbol{x}|C_k),$$

and the overall ECM is thus

$$\text{ECM} = \sum_{j=1}^{K} \text{ECM}(j) = \sum_{j=1}^{K} \sum_{k \neq j}^{K} c(k|j) p(k|j).$$

According to this criterion, a new observation $\boldsymbol{x}$ should be allocated by minimizing the expected cost of misclassification.

In the case of equal costs of misclassification ($c(k|j) = 1$ if $k \neq j$), we obtain

$$\text{ECM}(j) = \sum_{k \neq j}^{K} p(k|j) \propto \sum_{k \neq j}^{K} \tau_k f(\boldsymbol{x}|C_k),$$

and we allocate an observation $\boldsymbol{x}$ to the class C_k that minimizes $\text{ECM}(j)$, or, equivalently, that maximizes $\tau_k f(\boldsymbol{x}|C_k)$. This rule is the same as the standard MAP rule which uses the posterior probability from equation (4.2).

In the case of unequal costs of misclassification, consider the $K \times K$ matrix of costs having the form:

$$C = \{c(k|j)\} = \begin{bmatrix} 0 & c(2|1) & c(3|1) & \dots & c(K|1) \\ c(1|2) & 0 & c(3|2) & \dots & c(K|2) \\ c(1|3) & c(2|3) & 0 & \dots & c(K|3) \\ \vdots & \vdots & \ddots & \vdots & \vdots \\ c(1|K) & c(2|K) & c(3|K) & \dots & 0 \end{bmatrix}.$$

A simple solution can be obtained if we assume a constant cost of misclassifying an observation from class j, irrespective of the class predicted. Following Breiman et al. (1984, pp. 112–115), we can compute the per-class cost $c(k|j) = c(j)$ for $j \neq k$, and then get the predictions as in the unit-cost case with adjusted prior probabilities for the classes:

$$\tau_j^* = \frac{c(j)\tau_j}{\sum\limits_{j=1}^{K} c(j)\tau_j}.$$

Note that for two-class problems the use of the per-class cost vector is equivalent to using the original cost matrix.

EXAMPLE 4.4: Bankruptcy prediction based on financial ratios of corporations

Consider the data on financial ratios from Altman (1968), and available in the R package **MixGHD** (Tortora et al., 2022), which provides the ratio of retained earnings (RE) to total assets, and the ratio of earnings before interests and taxes (EBIT) to total assets, for 66 American corporations, of which half had filed for bankruptcy.

The following code loads the data and plots the financial ratios conditional on the class (see Figure 4.9):

```
data("bankruptcy", package = "MixGHD")
X <- bankruptcy[, -1]
Class <- factor(bankruptcy$Y, levels = c(1:0),
            labels = c("solvent", "bankrupt"))
cl <- clPairs(X, Class)
legend("bottomright", legend = cl$class,
     pch = cl$pch, col = cl $col, inset = 0.02)
```

Although the within-class distribution is clearly not Gaussian, in particular for the companies that have declared bankruptcy, we fit an EDDA classification model:

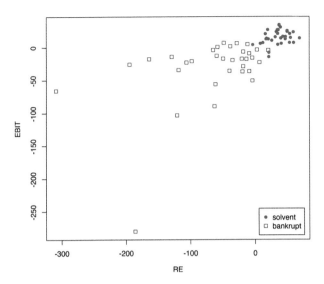

FIGURE 4.9: Scatterplot of financial ratios with points distinguished by observed classes.

```
mod <- MclustDA(X, Class, modelType = "EDDA")
summary(mod)
## ------------------------------------------------
## Gaussian finite mixture model for classification
## ------------------------------------------------
##
## EDDA model summary:
##
##  log-likelihood  n df      BIC
##         -661.27 66  8 -1356.1
##
## Classes      n  % Model G
##    solvent  33 50   VEE 1
##    bankrupt 33 50   VEE 1
##
## Training confusion matrix:
##             Predicted
## Class       solvent bankrupt
##    solvent       33        0
##    bankrupt       2       31
## Classification error = 0.0303
## Brier score          = 0.0295
```

The confusion matrix indicates that two training data points are misclassified.

Both are bankrupt firms which have been classified as solvent. The following plots show the distribution of financial ratios with (a) Gaussian ellipses implied by the estimated model, and (b) black points corresponding to the misclassified observations (see Figure 4.10):

```
plot(mod, what = "scatterplot")
plot(mod, what = "error")
```

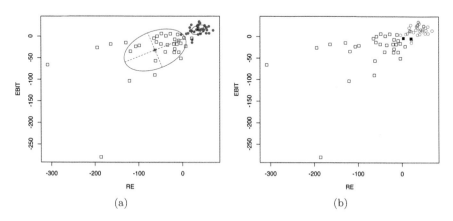

(a)   (b)

FIGURE 4.10: Scatterplots of financial ratios with points distinguished by observed classes. Panel (a) shows the ellipses implied by the estimated model. Panel (b) includes black points corresponding to the misclassified observations.

Now consider the error of misclassifying a bankrupt firm as solvent to be more serious than the opposite. We quantify these different costs of misclassification in a matrix C

```
(C <- matrix(c(0, 1, 10, 0), nrow = 2, ncol = 2, byrow = TRUE))
##      [,1] [,2]
## [1,]    0    1
## [2,]   10    0
```

and obtain the per-class cost vector:

```
rowSums(C)
## [1]  1 10
```

The total cost of misclassification for the MAP predictions is

```
pred <- predict(mod)
(tab <- table(Class, Predicted = pred$classification))
```

```
##              Predicted
## Class        solvent bankrupt
##    solvent      33        0
##    bankrupt      2       31
sum(tab * C)
## [1] 20
```

Unequal costs of misclassification can be included in the prediction as follows:

```
pred <- predict(mod, prop = mod$prop*rowSums(C))
(tab <- table(Class, Predicted = pred$classification))
##              Predicted
## Class        solvent bankrupt
##    solvent      28        5
##    bankrupt      0       33
sum(tab * C)
## [1] 5
```

The last command shows that we have been able to reduce the total cost by zeroing out the errors for bankrupt firms, at the same time increasing the errors for solvent corporations.

4.6   Classification with Unbalanced Classes

Most classification datasets are not balanced; that is, classes have unequal numbers of instances. Small differences between the number of instances in different classes can usually be ignored. In some cases, however, the imbalance in class proportions can be dramatic, and the class of interest is sometimes the class with the smallest number of cases. For instance, in studies aimed at identifying fraudulent transactions, classes are typically unbalanced, with the vast majority of the transactions not being fraudulent. In medical studies aimed at characterizing rare diseases, the class of individuals with the disease is only a small fraction of the total population (the *prevalence*).

In such situations, a case-control sampling scheme can be adopted by sampling approximately 50% of the data from the cases (fraudulent transactions, individuals suffering from a disease) and 50% from the controls. Balanced datasets can also be obtained in observational studies by *undersampling*, or downsizing the majority class by removing observations at random until the dataset is balanced. In both cases the class prior probabilities estimated from the training set do not reflect the "true" *a priori* probabilities. As a result, the predicted posterior class probabilities are not well estimated, resulting in a loss of classification accuracy compared to a classifier based on the true prior probabilities for the classes.

EXAMPLE 4.5: Classification of synthetic unbalanced two-class data

As an example, consider a simulated binary classification task, with the majority class having distribution $x|(y = 0) \sim N(0, 1)$, whereas the distribution of the minority class is $x|(y = 1) \sim N(3, 1)$, and in the population the latter accounts for 10% of cases. Suppose that a training sample is obtained using case-control sampling, so that the two groups have about the same proportion of cases.

A synthetic dataset from this specification can be simulated with the following code and shown graphically in Figure 4.11:

```
# generate training data from a balanced case-control sample
n_train <- 1000
class_train <- factor(sample(0:1, size = n_train, prob = c(0.5, 0.5),
                      replace = TRUE))
x_train <- ifelse(class_train == 1, rnorm(n_train, mean = 3, sd = 1),
                      rnorm(n_train, mean = 0, sd = 1))

hist(x_train[class_train == 0], breaks = 11, xlim = range(x_train),
     main = "", xlab = "x",
     col = adjustcolor("dodgerblue2", alpha.f = 0.5), border = "white")
hist(x_train[class_train == 1], breaks = 11, add = TRUE,
     col = adjustcolor("red3", alpha.f = 0.5), border = "white")
box()

# generate test data from mixture f(x) = 0.9 * N(0,1) + 0.1 * N(3,1)
n <- 10000
mixpro <- c(0.9, 0.1)
class_test <- factor(sample(0:1, size = n, prob = mixpro,
                     replace = TRUE))
x_test <- ifelse(class_test == 1, rnorm(n, mean = 3, sd = 1),
                     rnorm(n, mean = 0, sd = 1))
hist(x_test[class_test == 0], breaks = 15, xlim = range(x_test),
     main = "", xlab = "x",
     col = adjustcolor("dodgerblue2", alpha.f = 0.5), border = "white")
hist(x_test[class_test == 1], breaks = 11, add = TRUE,
     col = adjustcolor("red3", alpha.f = 0.5), border = "white")
box()
```

Using the training sample we can estimate a classification Gaussian mixture model:

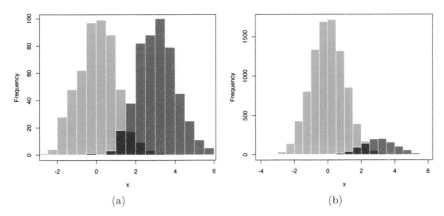

FIGURE 4.11: Histograms for synthetic datasets with observations sampled from two different Gaussian distributions. In the training data set, the cases $(y = 1)$ and the controls $(y = 0)$ are sampled in about the same proportions (a), whereas the cases $(y = 1)$ account for 10% of the observations in the whole population (b).

```
mod <- MclustDA(x_train, class_train)
summary(mod, parameters = TRUE)
## ------------------------------------------------
## Gaussian finite mixture model for classification
## ------------------------------------------------
##
## MclustDA model summary:
##
##  log-likelihood    n df     BIC
##         -1947.1 1000  4 -3921.9
##
## Classes    n      % Model G
##        0 505 50.5     X 1
##        1 495 49.5     X 1
##
## Class prior probabilities:
##     0     1
## 0.505 0.495
##
## Class = 0
##
## Mixing probabilities: 1
##
## Means:
```

```
## [1] 0.043067
##
## Variances:
## [1] 0.93808
##
## Class = 1
##
## Mixing probabilities: 1
##
## Means:
## [1] 3.0959
##
## Variances:
## [1] 0.96652
##
## Training confusion matrix:
##         Predicted
## Class    0    1
##     0  479   26
##     1   24  471
## Classification error = 0.05
## Brier score          = 0.0402
```

The estimated parameters for the class conditional distributions are close to the true values, but the prior class probabilities are highly biased due to the sampling scheme adopted. Performance measures can be computed for the test set:

```
pred <- predict(mod, newdata = x_test)
classError(pred$classification, class_test)$error
## [1] 0.0592
BrierScore(pred$z, class_test)
## [1] 0.045927
```

showing that they are slightly worse than those for the training set, as is often the case.

For such simulated data we know the true classes, so we can compute the performance measures over a grid of values of the prior probability of the minority class:

```
priorProp <- seq(0.01, 0.99, by = 0.01)
CE <- BS <- rep(as.double(NA), length(priorProp))
for (i in seq(priorProp))
{
  pred <- predict(mod, newdata = x_test,
```

```
                    prop = c(1-priorProp[i], priorProp[i]))
   CE[i] <- classError(pred$classification, class = class_test)$error
   BS[i] <- BrierScore(pred$z, class_test)
}
```

The following code produces Figure 4.12, which shows the classification error and the Brier score as functions of the prior probability for the minority class.

```
matplot(priorProp, cbind(CE, BS), type = "l", lty = 1, lwd = 2, xaxt = "n",
        xlab = "Class prior probability", ylab = "", ylim = c(0, max(CE, BS)),
        col = c("red3", "dodgerblue3"),
        panel.first =
            { abline(h = seq(0, 1, by = 0.05), col = "grey", lty = 3)
              abline(v = seq(0, 1, by = 0.05), col = "grey", lty = 3)
            })
axis(side = 1, at = seq(0, 1, by = 0.1))
abline(v = mod$prop[2],                # training proportions
       lty = 2, lwd = 2)
abline(v = mean(class_test == 1),   # test proportions (usually unknown)
       lty = 3, lwd = 2)
legend("topleft", legend = c("ClassError", "BrierScore"),
       col = c("red3", "dodgerblue3"), lty = 1, lwd = 2, inset = 0.02)
```

Vertical lines are drawn at the proportion for the minority class in the training set (dashed line), and at the proportion computed on the test set (dotted lines). However, the latter is usually unknown, but the plot clearly shows that there is room for improving the classification accuracy by using an unbiased estimate of the class prior probabilities.

To solve this problem, Saerens et al. (2002) proposed an EM algorithm that aims at estimating the adjusted posterior conditional probabilities of a classifier, and, as a by-product, provides estimates of the prior class probabilities. This method can be easily adapted to classifiers based on Gaussian mixtures.

Suppose a Gaussian mixture of the type in (4.1) is estimated on the training set $\mathcal{D}_{\text{train}} = \{(\boldsymbol{x}_1, y_1), \ldots, (\boldsymbol{x}_n, y_n)\}$, and a new test set $\mathcal{D}_{\text{test}} = \{\boldsymbol{x}_1^*, \ldots, \boldsymbol{x}_m^*\}$ is available. From equation (4.2) the posterior probabilities that an observation $\boldsymbol{x}_i^*$ belongs to class C_k are

$$\widehat{z}_{ik}^* = \widehat{\Pr}(C_k | \boldsymbol{x}_i^*) \quad \text{for } k = 1, \ldots, K.$$

Let $\tilde{\tau}_k = \sum_{i=1}^{n} I(y_i = C_k)/n$ be the proportion of cases from class k in the training set, and $\widehat{\tau}_k^0 = \sum_{i=1}^{m} \widehat{z}_{ik}^*/m$ be a preliminary estimate of the prior probabilities for class k ($k = 1, \ldots, K$). Starting with $s = 1$, the algorithm

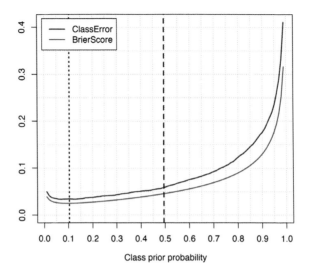

FIGURE 4.12: Classification error and Brier score as functions of the prior probability for the minority class. The vertical segments show the biased sample proportion of cases in the training set (dashed line), and the sample proportion of cases in the test set (dotted line), which is usually unknown.

iterates the following steps:

$$
\widehat{z}_{ik}^{(s)} = \frac{\dfrac{\widehat{\tau}_k^{(s-1)}}{\widetilde{\tau}_k}\widehat{z}_{ik}^{*}}{\displaystyle\sum_{g=1}^{K}\dfrac{\widehat{\tau}_g^{(s-1)}}{\widetilde{\tau}_k}\widehat{z}_{ig}^{*}} \;, \qquad\qquad \widehat{\tau}_k^{(s)} = \frac{1}{m}\sum_{i=1}^{m}\widehat{z}_{ik}^{(s)}\;,
$$

until the estimates $(\widehat{\tau}_1, \ldots, \widehat{\tau}_K)$ stabilize.

In **mclust** the estimated class prior probabilities, computed following the algorithm outlined above, can be obtained using the function `classPriorProbs()`.

EXAMPLE 4.6: Adjusting prior probabilities in unbalanced synthetic two-class data

Continuing the analysis in Example 4.5, the class prior probabilities are obtained as follows:

```
(priorProbs <- classPriorProbs(mod, x_test))
##        0       1
## 0.89452 0.10548
```

which provides estimates that are close to the true parameters. These can then be used to adjust the predictions to get:

```
pred <- predict(mod, newdata = x_test, prop = priorProbs)
classError(pred$classification, class = class_test)$error
## [1] 0.035
BrierScore(pred$z, class_test)
## [1] 0.025576
```

The performance measures can be contrasted with those obtained from the (usually unknown) class proportions in the test set:

```
(prior_test <- prop.table(table(class_test)))
## class_test
##     0     1
## 0.896 0.104
pred <- predict(mod, newdata = x_test, prop = prior_test)
classError(pred$classification, class = class_test)$error
## [1] 0.0351
BrierScore(pred$z, class_test)
## [1] 0.025568
```

4.7 Classification of Univariate Data

The classification of univariate data follows the same principles as discussed in previous sections, with the caveat that only two possible models are available in the EDDA case, namely E for equal within-class variance, and V for varying variances across classes. The same constraints on variances also apply to each mixture component within-class in MclustDA.

EXAMPLE 4.7: Classification of single-index linear combination for the Wisconsin diagnostic breast cancer data

To increase the accuracy of breast cancer diagnosis and prognosis, Mangasarian et al. (1995) used linear programming to identify the linear combination of features that optimally discriminate the benign from the malignant tumor cases. The linear combination was estimated to be

$$0.2322 \; \texttt{Texture\_mean} + 0.01117 \; \texttt{Area\_extreme} + 68.37 \; \texttt{Smoothness\_extreme}$$

With this extracted feature the authors reported being able to achieve a cross-validated predictive accuracy of 97.5% (which is equivalent to 0.025 misclassification error rate).

To illustrate the use of Gaussian mixtures to classify univariate data, we fit an MclustDA model to the same one-dimensional projection of the UCI wdbc data:

```
data("wdbc", package = "mclust")
x <- with(wdbc,
    0.2322*Texture_mean + 0.01117*Area_extreme + 68.37*Smoothness_extreme)
Class <- wdbc[, "Diagnosis"]
mod <- MclustDA(x, Class, modelType = "MclustDA")
summary(mod)
## -------------------------------------------------
## Gaussian finite mixture model for classification
## -------------------------------------------------
##
## MclustDA model summary:
##
##  log-likelihood    n df      BIC
##         -1763.1 569 10 -3589.7
##
## Classes    n      % Model G
##       B 357 62.74     X 1
##       M 212 37.26     V 3
##
## Training confusion matrix:
##      Predicted
## Class   B    M
##     B 351    6
##     M  14 198
## Classification error = 0.0351
## Brier score          = 0.0226
```

The selected model uses a single Gaussian component for the benign cancer
cases and a three-component heterogeneous mixture for the malignant cases.
The estimated within-class densities are shown in Figure 4.13, indicating a
bimodal distribution for the malignant tumors. This plot essentially agrees
with previous findings reported in Mangasarian et al. (1995, Fig. 3) using
non-parametric density estimation. The following code can be used to obtain
Figure 4.13:

```
(prop <- mod$prop)
##       B       M
## 0.62742 0.37258
col <- mclust.options("classPlotColors")
x0 <- seq(0, max(x)*1.1, length = 1000)
par1 <- mod$models[["B"]]$parameters
f1 <- dens(par1$variance$modelName, data = x0, parameters = par1)
par2 <- mod$models[["M"]]$parameters
f2 <- dens(par2$variance$modelName, data = x0, parameters = par2)
matplot(x0, cbind(prop[1]*f1, prop[2]*f2), type = "l", lty = 1,
```

```
        col = col, ylab = "Class density", xlab = "x")
legend("topright", title = "Diagnosis:", legend = names(prop),
        col = col, lty = 1, inset = 0.02)
```

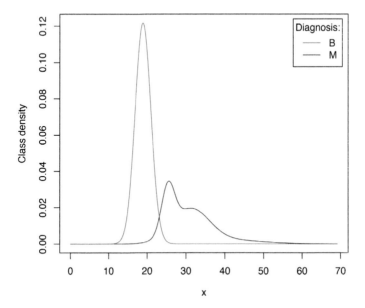

FIGURE 4.13: Densities for the benign and malignant tumors estimated using the univariate feature extracted from the breast cancer dataset.

The cross-validated misclassification error for the estimated MclustDA model is obtained as follows:

```
cv <- cvMclustDA(mod)  # by default: prop = mod$prop
unlist(cv[c("ce", "se.ce")])
##        ce       se.ce
## 0.0351494 0.0075474
```

Adjusting the class prior probabilities, as suggested by Mangasarian et al. (1995), we get:

```
cv <- cvMclustDA(mod, prop = c(0.5, 0.5))
unlist(cv[c("ce", "se.ce")])
##        ce       se.ce
## 0.0263620 0.0071121
```

Thus, assuming an equal class prior probability, the CV error rate goes down from 3.5% to 2.6%. It is possible to derive the corresponding discriminant thresholds numerically as:

```
x0 <- seq(min(x), max(x), length.out = 1000)
pred <- predict(mod, newdata = x0)
(threshold1 <- approx(pred$z[, 2], x0, xout = 0.5)$y)
## [1] 23.166
pred <- predict(mod, newdata = x0, prop = c(0.5, 0.5))
(threshold2 <- approx(pred$z[, 2], x0, xout = 0.5)$y)
## [1] 22.851
```

Note that the first threshold is equivalent to the point where the two densities in Figure 4.13 intersect. Using the default discriminant threshold, we can represent the training data graphically with the observed and predicted classes:

```
plot(mod, what = "scatterplot", main = TRUE)
abline(v = threshold1, lty = 2)
plot(mod, what = "classification", main = TRUE)
abline(v = threshold1, lty = 2)
```

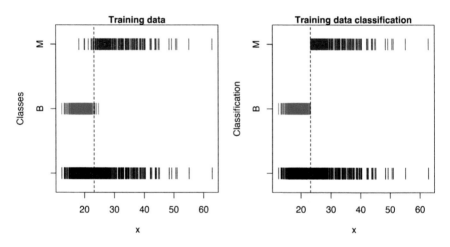

FIGURE 4.14: Distribution of the training data conditional on the true classes and on the predicted classes for the univariate feature extracted from the breast cancer data.

In the search for the optimal classification rule by cross-validation, we can proceed by tuning either the probability threshold for the decision rule or the prior class probabilities. The following code computes and plots the misclassification error rate as a function of the probability threshold used for the classification (see Figure 4.15):

```
threshold <- seq(0.1, 0.9, by = 0.05)
ngrid <- length(threshold)
cv <- data.frame(threshold, error = numeric(ngrid))
cverr <- cvMclustDA(mod, verbose = FALSE)
for (i in seq(threshold))
{
    cv$error[i] <- classError(ifelse(cverr$z[, 2] > threshold[i], "M", "B"),
                              Class)$errorRate
}
min(cv$error)
## [1] 0.024605
threshold[which.min(cv$error)]
## [1] 0.35
ggplot(cv, aes(x = threshold, y = error)) +
  geom_point() + geom_line() +
  scale_x_continuous(breaks = seq(0, 1, by = 0.1)) +
  ylab("CV misclassification error") +
  xlab("Probability threshold of malignant (M) tumor class")
```

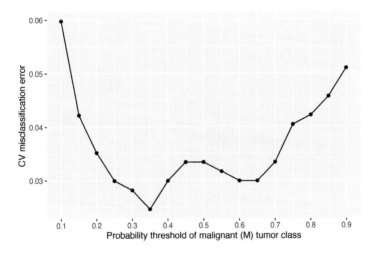

FIGURE 4.15: The cross-validated misclassification error rate as a function of the probability threshold for the univariate feature extracted from the breast cancer data.

It is thus possible to get a misclassification error of 2.46% by setting the probability threshold to 0.35.

A similar result can also be achieved by adjusting the class prior probabilities. However, because the CV procedure must be replicated for each value specified over a regular grid, this is more computationally demanding.

Nevertheless, the added computational effort has the advantage of providing an estimate of the standard error of the CV error estimate itself. This allows us to compute intervals around the CV estimate for assessing the significance of observed differences. The following code computes the CV misclassification error by adjusting the class prior probability of being a malignant tumor, followed by the code for plotting the CV results:

```
priorProb <- seq(0.1, 0.9, by = 0.05)
ngrid <- length(priorProb)
cv_error2 <- data.frame(priorProb,
                        cv = numeric(ngrid),
                        lower = numeric(ngrid),
                        upper = numeric(ngrid))
for (i in seq(priorProb))
{
  cv <- cvMclustDA(mod, prop = c(1-priorProb[i], priorProb[i]),
              verbose = FALSE)
  cv_error2$cv[i]    <- cv$ce
  cv_error2$lower[i] <- cv$ce - cv$se.ce
  cv_error2$upper[i] <- cv$ce + cv$se.ce
}
min(cv_error2$cv)
## [1] 0.026362
priorProb[which.min(cv_error2$cv)]
## [1] 0.5
ggplot(cv_error2, aes(x = priorProb, y = cv)) +
  geom_point() +
  geom_linerange(aes(ymin = lower, ymax = upper)) +
  scale_x_continuous(breaks = seq(0, 1, by = 0.1)) +
  ylab("CV misclassification error") +
  xlab("Malignant (M) tumor class prior probability")
```

By setting the class prior probabilities at 0.5 each, we get a misclassification error of 2.64%, which is similar to the result obtained by tuning the threshold. The strategy of assuming an equal prior class probability for each breast cancer class adopted by Mangasarian et al. (1995) appears to be the best approach available.

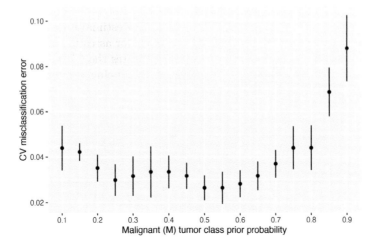

FIGURE 4.16: Plot of the CV misclassification error rate as a function of the class prior probability, with error bars shown at ± one standard error of the CV procedure, for the univariate feature extracted from the breast cancer data.

4.8 Semi-Supervised Classification

Supervised learning methods require knowing the correct class labels for the training data. However, in certain situations the available labeled data may be scarce because they are difficult or expensive to collect, for instance, in anomaly detection, computer-aided diagnosis, drug discovery, and speech recognition. Semi-supervised learning refers to models and algorithms that use both labeled and unlabeled data to perform certain learning tasks (Zhu and Goldberg, 2009; Van Engelen and Hoos, 2020). In classification scenarios, the main goal of semi-supervised learning is to train a classifier using both the labeled and unlabeled data such that its classification performance is better than the one obtained using a classifier trained on the labeled data alone.

Figure 4.17 shows a simulated two-class dataset with all the observations labeled (see left panel), and with only a few cases labeled (see right panel). In the typical supervised classification setting we would have access to the dataset in Figure 4.17a where labels are available for all the observations. When most of the labels are unknown, as in Figure 4.17b, two possible approaches are available. We could use only the (few) labeled cases to estimate a classifier, or we could use both labeled and unlabeled data in a semi-supervised learning approach. Figure 4.18 shows the classification boundaries corresponding to a supervised classification model estimated under the assumption that the true classes are known (solid line), the boundary for the classification model trained

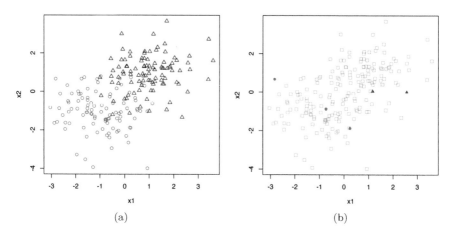

(a)   (b)

FIGURE 4.17: Example of two-class simulated dataset with (a) all labeled data (points are marked according to the true classes) and (b) only partial knowledge of labeled data (unlabeled data are shown as grey squares).

on the labeled data alone (dotted line), and the boundary obtained from a semi-supervised classification model (dashed line). The latter coincides almost exactly with the boundary arising from the full knowledge of class labels. In contrast, the classification boundary from the model estimated using only the labeled data is quite different.

Consider a training dataset $\mathcal{D}_{\text{train}}$ made of both l labeled cases $\{(\boldsymbol{x}_i, y_i)\}_{i=1}^{l}$ and u unlabeled cases $\{\boldsymbol{x}_j\}_{j=l+1}^{n}$, where $n = l + u$ is the overall sample size. As mentioned earlier, there are usually many more unlabeled than labeled data, so we assume that $u \gg l$. The likelihood of a semi-supervised mixture model depends on both the labeled and the unlabeled data, so the observed log-likelihood is

$$\ell(\boldsymbol{\Psi}) = \sum_{i=1}^{l}\sum_{k=1}^{K} c_{ik} \log\left\{\pi_k f_k(\boldsymbol{x}_i; \boldsymbol{\theta}_k)\right\} + \sum_{j=l+1}^{n} \log\left\{\sum_{g=1}^{G} \pi_g f_g(\boldsymbol{x}_j; \boldsymbol{\theta}_g)\right\},$$

where $c_{ik} = 1$ if observation i belongs to class k and 0 otherwise. Although in principle the number of components G can be greater than the number of classes K, it is usually assumed that $G = K$. If we treat the unknown labels as missing data, the complete-data log-likelihood can be written as

$$\ell_C(\boldsymbol{\Psi}) = \sum_{i=1}^{l}\sum_{k=1}^{K} c_{ik}\{\log(\pi_k) + \log(f_k(\boldsymbol{x}_i; \boldsymbol{\theta}_k))\}+$$

$$\sum_{j=l+1}^{n}\sum_{g=1}^{G} z_{jg}\{\log(\pi_g) + \log(f_g(\boldsymbol{x}_j; \boldsymbol{\theta}_g))\},$$

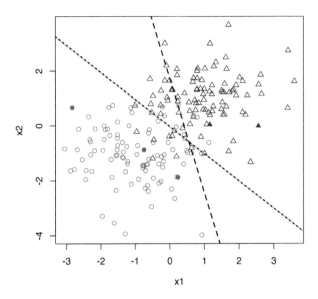

FIGURE 4.18: Classification boundaries for the two-class simulated dataset obtained (i) under the assumption of full knowledge of class labels (cyan solid line), (ii) using only the labeled data (black dashed line), and (iii) both labeled and unlabeled data (black dotted line).

where $z_{jg} \in (0, 1)$ is the conditional probability of observation j to belong to class g. By assuming a specific parametric distribution for the class and component densities, such as the multivariate Gaussian distribution, the EM algorithm can be used to find maximum likelihood estimates for the unknown parameters of the model. The estimated parameters are then used to classify the unlabeled data, as well as future data. More details can be found in McLachlan (1977), O'Neill (1978), McLachlan and Peel (2000, section 2.19) and Dean et al. (2006).

mclust provides an implementation for fitting Gaussian semi-supervised classification models through the function MclustSSC(). This requires the input of a data matrix and a vector of class labels, with unlabeled data encoded by NA. By default all the available models are fitted and the one with the largest BIC is returned. Optionally, the covariance decompositions described in Section 2.2.1 can be specified via the argument modelNames using the usual nomenclature from Table 2.1. In addition, the optional argument G can be used to specify the number of mixture components. By default, this is set equal to the number of classes from the labeled data.

EXAMPLE 4.8: Semi-supervised learning of Italian olive oils data

Consider the Italian olive oils dataset available in the **pgmm** R package

(McNicholas et al., 2022). The data are from a study conducted to determine the authenticity of olive oil (Forina et al., 1983) and provide the percentage composition of eight fatty acids found by lipid fraction of 572 Italian olive oils. The olive oils came from nine areas of Italy, which can be further grouped into three regions: Southern Italy, Sardinia, and Northern Italy. The following code reads the data, sets the data frame X of features to build the classifier, and creates the factor class containing the labels for all the observations:

```
data("olive", package = "pgmm")
X <- olive[, 3:10]
class <- factor(olive$Region, levels = 1:3,
                labels = c("South", "Sardinia", "North"))
table(class)
## class
##    South Sardinia    North
##      323       98      151
```

Knowing all the class labels, we can easily fit a discriminant analysis model using the EDDA mixture model discussed in Section 4.2:

```
mod_EDDA_full <- MclustDA(X, class, modelType = "EDDA")
pred_EDDA_full <- predict(mod_EDDA_full, newdata = X)
classError(pred_EDDA_full$classification, class)$errorRate
## [1] 0
BrierScore(pred_EDDA_full$z, class)
## [1] 0.00000010836
```

This model has very good performance, as measured by the classification error and the Brier score, although we must remember that this is an optimistic assessment of model performance because both metrics are computed on the data used for training. Despite this, we can use this performance as a benchmark.

Suppose that only partial knowledge of class labels is available. For instance, we can randomly retain only 10% of the labels as follows:

```
pct_labeled_data <- 10
n <- nrow(X)
cl <- class
is.na(cl) <- sample(1:n, round(n*(1-pct_labeled_data/100)))
table(cl, useNA = "ifany")
## cl
##    South Sardinia    North     <NA>
##       28       11       18      515
```

Since approximately 90% of the data are unlabeled, a semi-supervised classification is called for:

```
mod_SSC <- MclustSSC(X, cl)
plot(mod_SSC, what = "BIC")
```

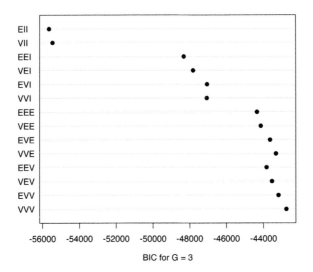

FIGURE 4.19: BIC values for the semi-supervised classification models fitted to the Italian olive oils data using 10% of labeled data.

```
mod_SSC$BIC
##       EII      VII      EEI      VEI      EVI      VVI      EEE      VEE      EVE      VVE
## 3 -55641 -55449 -48321 -47813 -47057 -47073 -44348 -44140 -43640 -43323
##       EEV      VEV      EVV      VVV
## 3 -43829 -43539 -43191 -42768
pickBIC(mod_SSC$BIC, 5) - max(mod_SSC$BIC)  # BIC diff for the top-5 models
##    VVV,3    EVV,3    VVE,3    VEV,3    EVE,3
##     0.00 -422.20 -554.53 -770.71 -871.55
```

Figure 4.19 plots the BIC values for the models fitted by `MclustSSC()` with $G = K = 3$ classes. The best estimated model according to BIC is the unconstrained VVV model. The `summary()` function can be used to obtain a summary of the fit:

```
summary(mod_SSC)
## ------------------------------------------------------------------
## Gaussian finite mixture model for semi-supervised classification
## ------------------------------------------------------------------
##
##  log-likelihood   n   df     BIC
```

```
##             -20959 572 134 -42768
##
## Classes     n    % Model G
##    South    28  4.90   VVV 1
##    Sardinia 11  1.92   VVV 1
##    North    18  3.15   VVV 1
##    <NA>    515 90.03
##
## Classification summary:
##               Predicted
## Class      South Sardinia North
##    South      28        0     0
##    Sardinia    0       11     0
##    North       0        0    18
##    <NA>      295       86   134
```

We can evaluate the estimated classifier by comparing the predicted classes with the true classes for the unlabeled observations:

```
pred_SSC <- predict(mod_SSC, newdata = X[is.na(cl), ])
table(Predicted = pred_SSC$classification, Class = class[is.na(cl)])
##            Class
## Predicted  South Sardinia North
##    South      295        0     0
##    Sardinia     0       86     0
##    North        0        1   133
classError(pred_SSC$classification, class[is.na(cl)])$errorRate
## [1] 0.0019417
BrierScore(pred_SSC$z, class[is.na(cl)])
## [1] 0.0019417
```

The performance appears quite good with only one unlabeled observation misclassified.

We now compare the semi-supervised approach with the classification approach that uses only the labeled data, as a function of the percentage of full data information. The following code implements this comparison using the Brier score as performance metric:

```
pct_labeled_data <- c(5, seq(10, 90, by = 10), 95)
BS <- matrix(as.double(NA), nrow = length(pct_labeled_data), ncol = 2,
             dimnames = list(pct_labeled_data, c("EDDA", "SSC")))
for (i in seq(pct_labeled_data))
{
  cl <- class
  labeled <- sample(1:n, round(n*pct_labeled_data[i]/100))
```

```
   cl[-labeled] <- NA
   # Classification on labeled data
   mod_EDDA  <- MclustDA(X[labeled, ], cl[labeled],
                         modelType = "EDDA")
   # prediction for the unlabeled data
   pred_EDDA <- predict(mod_EDDA, newdata = X[-labeled, ])
   BS[i, 1]  <- BrierScore(pred_EDDA$z, class[-labeled])
   # Semi-supervised classification
   mod_SSC  <- MclustSSC(X, cl)
   # prediction for the unlabeled data
   pred_SSC <- predict(mod_SSC, newdata = X[-labeled, ])
   BS[i, 2] <- BrierScore(pred_SSC$z, class[-labeled])
}
BS
##                     EDDA                  SSC
## 5   0.36753964900313729 0.00184075724060626
## 10  0.14649511848830160 0.00194083725139750
## 20  0.00000004794698661 0.00218238248453443
## 30  0.00500200204812944 0.0000000044264099
## 40  0.00356335895644646 0.00000009331253357
## 50  0.01740046202639491 0.00349653566200317
## 60  0.00000033388757151 0.00000014123579678
## 70  0.00585344025094451 0.00000037530321144
## 80  0.00002069320555273 0.00000027964277368
## 90  0.00000000012729332 0.00000000000499843
## 95  0.00000000000069483 0.00000000000052263
matplot(pct_labeled_data, BS, type = "b",
        lty = 1, pch = c(19, 15), col = c(2, 4), xaxt = "n",
        xlab = "Percentage of labeled data", ylab = "Brier score")
axis(side = 1, at = pct_labeled_data)
abline(h = BrierScore(pred_EDDA_full$z, class), lty = 2)
legend("topright", pch = c(19, 15), col = c(2, 4), lty = 1,
        legend = c("EDDA", "SSC"), inset = 0.02)
```

Looking at the table of results and Figure 4.20, it is clear that EDDA requires a large percentage of labeled data to achieve a reasonable performance. In contrast, the semi-supervised classification (SSC) model is able to achieve almost perfect classification accuracy with a very small portion of labeled data.

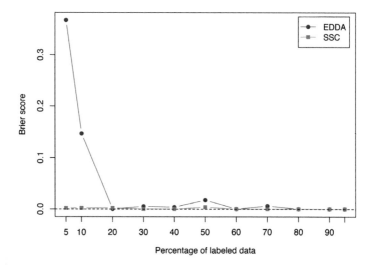

FIGURE 4.20: Brier score values for the EDDA classification model on labeled data and the semi-supervised classification (SSC) model as a function of the percentage of labeled data.

5

Model-Based Density Estimation

Density estimation plays an important role in both applied data analysis and theoretical research in statistics. Finite mixture models provide a flexible semiparametric methodology for density estimation, where the unknown density can be regarded as a convex linear combination of one or more probability density functions. This chapter illustrates data examples where Gaussian mixtures are used for model-based univariate and multivariate density estimation.

5.1 Density Estimation

Density estimation is a valuable tool for summarizing data distributions. A good density estimate can reveal important characteristics of the data as well as provide a useful exploratory tool. Broadly speaking, three alternative approaches can be distinguished: *parametric* density estimation, *nonparametric* density estimation, and *semiparametric* density estimation via finite mixture modeling.

In the parametric approach, a distribution is assumed for the density with unknown parameters that are estimated by fitting a parametric function to the observed data. In the nonparametric approach, no density function is assumed a priori; rather, its form is completely determined by the data. Histograms and kernel density estimation (KDE) are two popular methods that belong to this class, and in both the number of parameters generally grows with the sample size and dimensionality of the dataset (Silverman, 1998; Scott, 2009). For univariate data, functions hist() and density() are available in base R for nonparametric density estimation, and several other packages, such as **KernSmooth** (Wand, 2021), **ks** (Duong, 2022), and **sm** (Bowman et al., 2022), either include functionality or are devoted to this approach. However, extension to higher dimensions is less well established.

Finite mixture models provide a flexible semiparametric methodology for density estimation. In this approach, the unknown density is expressed as a convex linear combination of one or more probability density functions. The Gaussian mixture model (GMM), which assumes the Gaussian distribution for the underlying component densities, is a popular choice in this class of methods. GMMs can approximate any continuous density with arbitrary accuracy, provided the model has a sufficient number of components (Ferguson, 1983; Marron and Wand, 1992; Escobar and West, 1995; Roeder and Wasserman, 1997; Frühwirth-Schnatter, 2006). The number of parameters increases with the number of components, as well as with the dimensionality of the data except in the simplest case.

Advantages of the Gaussian mixture modeling in this context include:

• No need to specify tuning parameters, such as the number of bins and the origin for histograms, or the bandwidth for kernel density estimation;

• Efficient even for multidimensional data;

• Maximum likelihood estimates are available.

Disadvantages of the Gaussian mixture modeling approach include:

• Estimation via the EM algorithm may be slow and requires good starting values;

• Bias-variance trade-off: many GMM components may be needed for approximating a distribution, increasing the variance of the estimates, while more parsimonious models reduce the variability but may introduce some bias.

• The number of components to include is not known in advance.

5.2   Finite Mixture Modeling for Density Estimation with mclust

Consider a vector of random variables x taking values in the sample space $\mathbb{R}^d$ with $d \geq 1$, and assume that the probability density function can be written as a finite mixture density of G components as in equation (2.1). The model adopted in this chapter assumes a Gaussian distribution for each component density, so the density of x can be written as

$$f(x) = \sum_{k=1}^{G} \pi_k \phi(x; \mu_k, \Sigma_k),$$

where $\phi(\cdot)$ is the (multivariate) Gaussian density with mean μ_k, covariance matrix Σ_k, and mixing weight π_k for component k ($\pi_k > 0$, $\sum_{k=1}^{G} \pi_k = 1$), with $k = 1, \ldots, G$. This is essentially a GMM of the form given in equation 2.3.

Gaussian finite mixture modeling is a general strategy for density estimation. Nonparametric kernel density estimation (KDE) can be viewed as a mixture of $G = n$ components with uniform weights: $\pi_k = 1/n$ (Titterington et al., 1985, pages 28–29). Compared to KDE, finite mixture modeling typically uses a smaller number of components and hence fewer parameters. Conversely, compared to fully parametric density estimation, finite mixture modeling has the potential advantage of using more parameters and so introducing less estimation bias.

Mixture modeling also has its drawbacks, such as increased learning complexity and the need for iterative numerical procedures (such as the EM algorithm) for estimation. In certain cases there can also be identifiability issues (see Section 2.1.2).

mclust provides a simple interface to Gaussian mixture models for univariate and multivariate density estimation through the densityMclust() function. Available arguments and functionalities are analogous to those described in Section 3.2 for the Mclust() function in the clustering case. Several examples of its usage are provided in the following sections.

5.3 Univariate Density Estimation

EXAMPLE 5.1: Density estimation of Hidalgo stamp data

Izenman and Sommer (1988) considered fitting a Gaussian mixture to the distribution of the thickness of stamps in the 1872 Hidalgo stamp issue of Mexico. The dataset is available in R package **multimode** (Ameijeiras-Alonso et al., 2021).

```
data("stamps1", package = "multimode")
str(stamps1)
## 'data.frame': 485 obs. of  2 variables:
## $ thickness: num 0.06 0.064 0.064 0.065 0.066 0.068 0.069 0.069 0.069
##    0.069 ...
## $ year : Factor w/ 2 levels "1872","1873-1874": 2 2 2 2 2 1 1 1 1 2 ...
Thickness <- stamps1$thickness
```

A density estimate based on GMMs can be obtained in **mclust** with the function densityMclust(), which also produces a plot of the estimated density:

```
dens <- densityMclust(Thickness)
```

The plot can be suppressed by setting the optional argument plot = FALSE. A summary can be obtained as follows:

```
summary(dens, parameters = TRUE)
## ------------------------------------------------------------
```

```
## Density estimation via Gaussian finite mixture modeling
## ----------------------------------------------------------
##
## Mclust V (univariate, unequal variance) model with 3 components:
##
##  log-likelihood    n df    BIC    ICL
##          1516.6  485  8 2983.8 2890.9
##
## Mixing probabilities:
##        1       2       3
## 0.26535 0.30179 0.43286
##
## Means:
##        1        2        3
## 0.072145 0.079346 0.099189
##
## Variances:
##             1            2            3
## 0.0000047749 0.0000031193 0.0001886446
```

This summary output shows that the estimated model selected by BIC is a three-component mixture with different variances (V,3), and gives the values of the estimated parameters.

The density estimate can also be displayed with the associated plot() method. For instance, Figure 5.1 shows the estimated density drawn over a histogram of the observed data. The latter is added by providing the optional argument data and specifying the break points between histogram bars (in the base R hist() function) by setting the argument breaks:

```
br <- seq(min(Thickness), max(Thickness), length.out = 21)
plot(dens, what = "density", data = Thickness, breaks = br)
```

Three modes are clearly visible in the plot. These appear at the values of the mean of each mixture component: one component with larger mean and dispersion of stamp thickness, and two components having thinner and less variable stamps:

```
with(dens$parameters,
     data.frame(mean = mean,
                sd = sqrt(variance$sigmasq),
                CoefVar = sqrt(variance$sigmasq)/mean*100))
##       mean        sd CoefVar
## 1 0.072145 0.0021852  3.0288
## 2 0.079346 0.0017662  2.2259
## 3 0.099189 0.0137348 13.8470
```

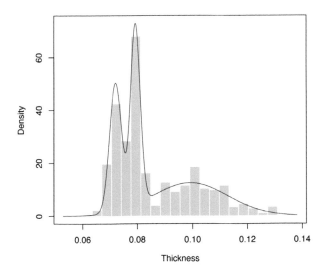

FIGURE 5.1: Histogram of thickness for the Hidalgo 1872 stamps1 dataset, with the GMM density estimate superimposed.

A predict() method is associated with objects of class "densityMclust", for computing either the overall density (what = "dens", the default), or the individual (unweighted) mixture component densities (what = "cdens") at specified data points. Densities are returned on the logarithmic scale if we set logarithm = TRUE. For instance:

```
x <- c(0.07, 0.08, 0.1, 0.12)
predict(dens, newdata = x, what = "dens")
## [1] 31.2340 68.4716 12.5509  3.9895
predict(dens, newdata = x, logarithm = TRUE)
## [1] 3.4415 4.2264 2.5298 1.3837
predict(dens, newdata = x, what = "cdens")
##                1             2        3
## [1,]  1.1275e+02  1.8734e-04   3.0362
## [2,]  2.8553e-01  2.1093e+02  10.9450
## [3,]  9.4735e-34  4.5561e-28  28.9956
## [4,] 1.3055e-102 2.0026e-113   9.2166
```

The matrix of posterior conditional probabilities is obtained by specifying what = "z":

```
predict(dens, newdata = x, what = "z")
##               1          2        3
## [1,]  9.5792e-01 1.8102e-06 0.042077
## [2,]  1.1065e-03 9.2970e-01 0.069191
```

```
## [3,]  2.0029e-35  1.0955e-29 1.000000
## [4,] 8.6832e-104 1.5149e-114 1.000000
```

The stamps1 dataset contains additional information that can be used to shed
light on the selected model. In particular, thickness measurements can be
grouped according to the year of consignment: the first 289 stamps refer to
the 1872 issue and the remaining 196 stamps to the years 1873–1874. We may
draw a (suitably scaled) histogram for each year-of-consignment and then add
the estimated component densities as follows:

```
Year <- stamps1$year
table(Year)
## Year
##      1872 1873-1874
##       289       196
h1 <- hist(Thickness[Year == "1872"], breaks = br, plot = FALSE)
h1$density <- h1$density*prop.table(table(Year))[1]
h2 <- hist(Thickness[Year == "1873-1874"], breaks = br, plot = FALSE)
h2$density <- h2$density*prop.table(table(Year))[2]
x <- seq(min(Thickness)-diff(range(Thickness))/10,
         max(Thickness)+diff(range(Thickness))/10, length = 200)
cdens <- predict(dens, x, what = "cdens")
cdens <- sweep(cdens, 2, dens$parameters$pro, "*")
col <- adjustcolor(mclust.options("classPlotColors")[1:2], alpha = 0.3)
ylim <- range(h1$density, h2$density, cdens)
plot(h1, xlab = "Thickness", freq = FALSE, main = "", border = "white",
col = col[1], xlim = range(x), ylim = ylim)
plot(h2, add = TRUE, freq = FALSE, border = "white", col = col[2])
matplot(x, cdens, type = "l", lwd = 1, lty = 1, col = 1, add = TRUE)
box()
legend("topright", legend = levels(Year), col = col, pch = 15, inset = 0.02,
       title = "Overprinted years:", title.adj = 0.2)
```

The result is shown in Figure 5.2. Stamps from 1872 show a two-part distribution, with one component corresponding to the largest thickness, and one
whose distribution essentially overlaps with the bimodal distribution of stamps
for the years 1873–1874.

EXAMPLE 5.2: Density estimation of acidity data

Consider the distribution of an acidity index measured on a sample of 155
lakes in the Northeastern United States. The data have been analyzed on
the logarithmic scale using mixtures of normal distributions by Richardson
and Green (1997) and McLachlan and Peel (2000, Sec. 6.6.2). The dataset
is included in the CRAN package **BNPdensity** (Arbel et al., 2021; Barrios
et al., 2021) and can be accessed as follows:

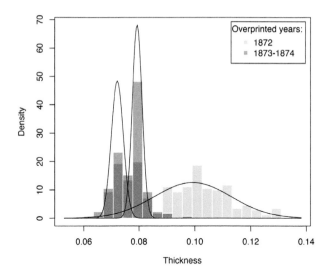

FIGURE 5.2: Histograms of thickness by overprinted year for the Hidalgo 1872 stamps1 dataset, with mixture component-density estimates superimposed.

```
data("acidity", package = "BNPdensity")
```

Using BIC as the model selection criterion, the "best" estimated model with respect to both the variance structure and the number of components can be obtained as follows:

```
summary(mclustBIC(acidity), k = 5)
## Best BIC values:
##               E,2        V,3        V,2        E,5       E,3
## BIC       -392.07 -397.9108 -399.6945 -400.6745 -402.183
## BIC diff    0.00    -5.8385    -7.6222    -8.6022   -10.111
```

BIC supports the two-component model with equal variance across the components (E,2). This model can be estimated as follows:

```
dens_E2 <- densityMclust(acidity, G = 2, modelNames = "E")
summary(dens_E2, parameters = TRUE)
## --------------------------------------------------------
## Density estimation via Gaussian finite mixture modeling
## --------------------------------------------------------
##
## Mclust E (univariate, equal variance) model with 2 components:
##
## log-likelihood  n df     BIC     ICL
```

```
##            -185.95 155   4 -392.07 -398.56
##
## Mixing probabilities:
##         1        2
## 0.62336 0.37664
##
## Means:
##        1      2
## 4.3709 6.3202
##
## Variances:
##         1        2
## 0.18637 0.18637
```

Next, we provisionally assume a GMM with unequal variances and decide the number of mixture components via the bootstrap LRT discussed in Section 2.3.2 instead of the BIC:

```
mclustBootstrapLRT(acidity, modelName = "V")
## ----------------------------------------------------------------
## Bootstrap sequential LRT for the number of mixture components
## ----------------------------------------------------------------
## Model        = V
## Replications = 999
##               LRTS bootstrap p-value
## 1 vs 2    77.0933            0.001
## 2 vs 3    16.9140            0.004
## 3 vs 4     5.1838            0.286
```

LRT suggests a three-component Gaussian mixture (V,3), which was a fairly close second-best fit according to BIC. This (V,3) model can be estimated as follows:

```
dens_V3 <- densityMclust(acidity, G = 3, modelNames = "V")
summary(dens_V3, parameters = TRUE)
## ------------------------------------------------------------
## Density estimation via Gaussian finite mixture modeling
## ------------------------------------------------------------
##
## Mclust V (univariate, unequal variance) model with 3 components:
##
##   log-likelihood   n df     BIC      ICL
##          -178.78 155  8 -397.91 -458.86
##
## Mixing probabilities:
```

```
##         1       2       3
## 0.34061 0.31406 0.34534
##
## Means:
##        1      2      3
## 4.2040 4.6796 6.3809
##
## Variances:
##          1        2        3
## 0.044195 0.338309 0.178276
```

The parameter estimates for the GMMs are available from the `summary()` output, and the corresponding density functions can be plotted with the following code:

```
plot(dens_E2, what = "density",
     ylim = c(0, max(dens_E2$density, dens_V3$density)))
rug(acidity)
plot(dens_V3, what = "density",
     ylim = c(0, max(dens_E2$density, dens_V3$density)))
rug(acidity)
```

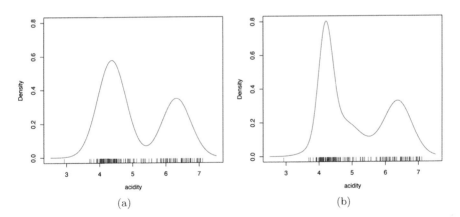

(a)   (b)

FIGURE 5.3: Density estimates provided by (a) the model (E,2) supported by BIC and (b) the model (V,3) supported by the LRT for the `acidity` data.

Figure 5.3 shows the plots of the estimated densities. The two models largely agree, with the exception of a somewhat different shape for some lower-range values. In particular, both densities indicate a bimodal distribution.

5.3.1 Diagnostics for Univariate Density Estimation

Two diagnostic plots for density estimation are available for univariate data (for details see Loader, 1999, pages 87–90). The first plot is a graph of the estimated cumulative distribution function (CDF) against the empirical distribution function. The second is a Q-Q plot of the sample quantiles against the quantiles computed from the estimated density. For a GMM, the estimated CDF is just the weighted sum of individual CDFs for each component. It can be obtained in **mclust** via the function cdfMclust().

EXAMPLE 5.3: Diagnostic for density estimation of acidity data

Recalling Example 5.2 on acidity data, diagnostic plots for the (E,2) model can be obtained as follows:

```
plot(dens_E2, what = "diagnostic", type = "cdf")
plot(dens_E2, what = "diagnostic", type = "qq")
```

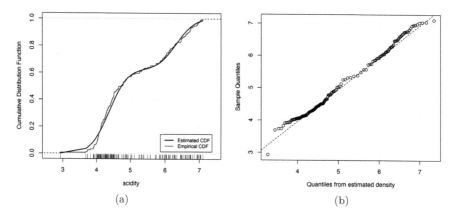

(a)   (b)

FIGURE 5.4: Density estimate diagnostics for the (E,2) model estimated for the acidity data: (a) estimated CDF and empirical distribution function, (b) Q-Q plot of sample quantiles vs. quantiles from density estimation.

Similarly, for the model (V,3):

```
plot(dens_V3, what = "diagnostic", type = "cdf")
plot(dens_V3, what = "diagnostic", type = "qq")
```

Figure 5.4a and Figure 5.5a compare the estimated CDFs to the empirical CDF. Figure 5.4b and Figure 5.5b show the corresponding Q-Q plots for the two models. Both types of diagnostics seem to suggest a better fit to the

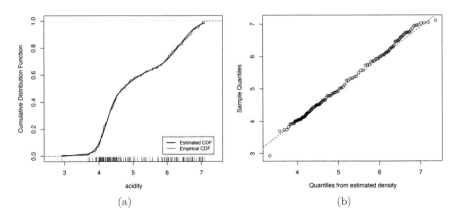

FIGURE 5.5: Density estimate diagnostics for the (V,3) model estimated for the `acidity` data: (a) estimated CDF and empirical distribution function, (b) Q-Q plot of sample quantiles vs. quantiles from density estimation.

observed data for the more complex model, with perhaps one outlying point corresponding to the smallest value.

5.4 Density Estimation in Higher Dimensions

In principle, density estimation in higher dimensions is straightforward with **mclust**. The structure imposed by the Gaussian components allows a parsimonious representation of the underlying density, especially when compared to KDE approaches. However, density estimation in high dimensions is problematic in general. Another issue is that visualizing densities beyond three dimensions is difficult. Following the suggestion of Scott (2009, p. 217), we should investigate "density estimation in several dimensions rather than in very high dimensions."

EXAMPLE 5.4: Density estimation of Old Faithful data

Consider the Old Faithful data described in Example 3.4. A scatterplot of the data is shown in Figure 5.6a. This and the corresponding density estimate can be obtained with the following code:

```
data("faithful", package = "datasets")
plot(faithful)
dens <- densityMclust(faithful)
summary(dens, parameters = TRUE)
## ----------------------------------------------------------
## Density estimation via Gaussian finite mixture modeling
## ----------------------------------------------------------
##
## Mclust EEE (ellipsoidal, equal volume, shape and orientation) model
## with 3 components:
##
## log-likelihood   n df      BIC      ICL
##        -1126.3 272 11 -2314.3 -2357.8
##
## Mixing probabilities:
##       1       2       3
## 0.16568 0.35637 0.47795
##
## Means:
##               [,1]    [,2]    [,3]
## eruptions   3.7931  2.0376  4.4632
## waiting    77.5211 54.4912 80.8334
##
## Variances:
## [,,1]
##           eruptions waiting
## eruptions  0.078254  0.4802
```

```
## waiting     0.480198 33.7671
## [,,2]
##              eruptions waiting
## eruptions   0.078254  0.4802
## waiting     0.480198 33.7671
## [,,3]
##              eruptions waiting
## eruptions   0.078254  0.4802
## waiting     0.480198 33.7671
```

Model selection based on the BIC selects a three-component mixture with common covariance matrix (EEE,3). One component is used to model the group of observations having both low duration and low waiting times, whereas two components are needed to approximate the skewed distribution of the observations with larger duration and waiting times.

For two-dimensional data, the plot()mclust!plot.densityMclust() method for the argument what = "density" is by default a contour plot of the density estimate, as shown in Figure 5.6b:

```
plot(dens, what = "density")
```

A bivariate density estimate may also be plotted using an image plot or a perspective plot (see Figures 5.6c–5.6d):

```
plot(dens, what = "density", type = "image")
plot(dens, what = "density", type = "persp")
```

In the above code we specified the type of density plot to produce. Several optional arguments for customizing the plots are available; for a complete description see help("plot.densityMclust") and help("surfacePlot"). More details on producing graphical displays with **mclust** are discussed in Sections 6.2 and 6.3.

Another useful approach to visualizing multivariate densities involves summarizing a density estimate with (possibly disjoint) regions of the sample space covering a specified probability. These are called Highest Density Regions (HDRs) by Hyndman (1996); for details on the definition and the computational implementation see Section 5.6. In **mclust**, HDRs can be easily obtained by specifying type = "hdr" in the plot() function applied to an object returned by a densityMclust() function call:

```
plot(dens, what = "density", type = "hdr")
```

For higher dimensional datasets densityMclust() provides graphical displays

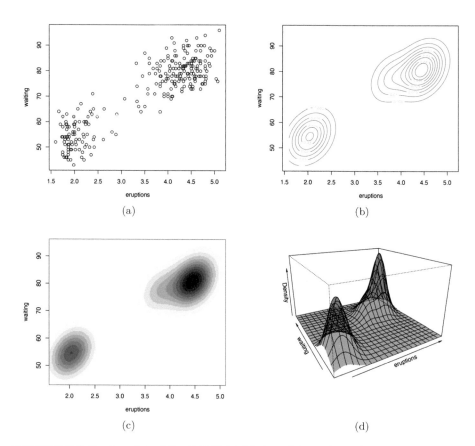

FIGURE 5.6: Scatterplot of the Old Faithful data (a), **mclust** density estimate represented as contour levels (b), image plot of density estimate (c), and perspective plot of the bivariate density estimate (d).

of density estimates using density contour, image, and perspective plots for pairs of variables arranged in a matrix.

EXAMPLE 5.5: Density estimation on principal components of the aircraft data

Consider the dataset aircraft, available in the **sm** package (Bowman et al., 2022) for R, giving measurements on six physical characteristics of twentieth-century aircraft. Following the analysis of Bowman and Azzalini (1997, Sec. 1.3), we consider the principal components computed from the data on the log scale for the time period 1956–1984:

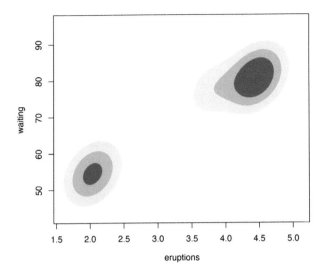

FIGURE 5.7: Highest density regions from the density estimated on the faithful data at probability levels 0.25, 0.5, and 0.75.

```
data("aircraft", package = "sm")
X <- log(subset(aircraft, subset = (Period == 3), select = 3:8))
PCA <- prcomp(X, scale = TRUE)
summary(PCA)
## Importance of components:
##                          PC1   PC2    PC3     PC4     PC5     PC6
## Standard deviation     2.085 1.064 0.6790 0.16892 0.15239 0.08771
## Proportion of Variance 0.725 0.189 0.0769 0.00476 0.00387 0.00128
## Cumulative Proportion  0.725 0.913 0.9901 0.99485 0.99872 1.00000
PCA$rotation        # loadings
##             PC1      PC2      PC3        PC4       PC5       PC6
## Power   -0.45612  0.262445 -0.11357  0.2267760 -0.565774 -0.581940
## Span    -0.37433 -0.540127 -0.32461  0.5305967  0.417040 -0.085530
## Length  -0.46737 -0.088898 -0.21657 -0.7885789  0.260671 -0.192237
## Weight  -0.47409  0.036498 -0.17017  0.0121226 -0.365831  0.781647
## Speed   -0.29394  0.731054  0.15713  0.2121675  0.551014  0.076387
## Range   -0.34963 -0.309371  0.88384 -0.0045952 -0.024057 -0.016386
Z <- PCA$x[, 1:3]    # PCA projection
```

Consider the fit of a GMM on the first three principal components (PCs), which explain 99% of total variability. The BIC criterion indicates a number of candidate models with almost the same support from the data.

```
BIC <- mclustBIC(Z)
summary(BIC, k = 5)
## Best BIC values:
##              VEE,5         VVE,4        VEI,6       VVE,3       VEE,4
## BIC      -1958.7 -1958.752723 -1961.8724 -1962.876 -1965.7162
## BIC diff    0.0     -0.014737    -3.1344    -4.138    -6.9782
```

We fit the simpler three-component model with varying volume and shape
(VVE), and plot the estimated density (see Figure 5.8).

```
densAircraft <- densityMclust(Z, G = 3, modelNames = "VVE", plot = FALSE)
plot(densAircraft, what = "density", type = "hdr",
     data = Z, points.cex = 0.5)
```

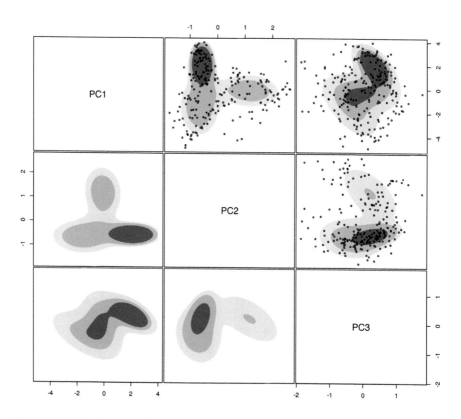

FIGURE 5.8: Scatterplot matrix of selected principal components of the
aircraft data with bivariate highest density regions at probability levels 0.25,
0.5, and 0.75.

The default probability levels 0.25, 0.5, and 0.75 are used in these plots (for details see Section 5.6). The plots are arranged in a symmetric matrix, where each off-diagonal panel shows the bivariate marginal density estimate. Data points have been also added to the upper-diagonal panels by providing the matrix of principal components to the optional argument data. The presence of several modes is clearly evident from this figure.

5.5 Density Estimation for Bounded Data

Finite mixtures of Gaussian distributions provide a flexible model for density estimation when the continuous variables under investigation have no boundaries. However, in practical applications variables may be partially bounded (taking non-negative values) or completely bounded (taking values in the unit interval). In this case, the standard Gaussian finite mixture model assigns non-zero densities to any possible value, even to those outside the ranges where the variables are defined, hence resulting in potentially severe bias.

A transformation-based approach for Gaussian mixture modeling in the case of bounded variables has been recently proposed (Scrucca, 2019). The basic idea is to carry out density estimation not on the original data $\boldsymbol{x}$ with bounded support $\mathcal{S}_{\mathcal{X}} \subset \mathbb{R}^d$, but on the transformed data $\boldsymbol{y} = t(\boldsymbol{x}; \boldsymbol{\lambda})$ having unbounded support $\mathcal{S}_{\mathcal{Y}}$. A *range-power transformation* is used for $t(\cdot; \boldsymbol{\lambda})$. This is a monotonic transformation which depends on parameters $\boldsymbol{\lambda}$ and maps $\mathcal{S}_{\mathcal{X}}$ to $\mathcal{S}_{\mathcal{Y}}$. The range part of the transformation differs depending on whether a variable has only a lower bound or has both a lower and an upper bound, while the power part of the transformation involves the well-known Box-Cox transformation (Box and Cox, 1964). Once the density on the transformed scale is estimated, the density for the original data can be obtained by a change of variables using the Jacobian of the transformation. Both the transformation parameters and the parameters of the Gaussian mixture are jointly estimated by the EM algorithm. For more details see Scrucca (2019).

The methodology briefly outlined above is implemented in the R package **mclustAddons** (Scrucca, 2022):

```
library("mclustAddons")
```

EXAMPLE 5.6: Density estimation of suicide data

The dataset gives the lengths (in days) of 86 spells of psychiatric treatment undergone by control patients in a suicide risk study (Silverman, 1998). Since the variable can only take positive values, in the following we contrast the density estimated ignoring this fact with the estimate obtained using the approach outlined above.

The following code estimates the density using the function densityMclust() with its default settings:

```
data("suicide", package = "mclustAddons")
dens <- densityMclust(suicide)
rug(suicide)              # add data points at the bottom of the graph
abline(v = 0, lty = 3)  # draw a vertical line at the natural boundary
```

The resulting density is shown in Figure 5.9a. This is clearly unsatisfactory because it assigns nonzero density to negative values and exhibits a bimodal distribution, the latter being an artifact due to failure to account for the non-negativity.

The function densityMclustBounded() can be used for improved density estimation by specifying the lower bound of the variable:

```
bdens <- densityMclustBounded(suicide, lbound = 0)
summary(bdens, parameters = TRUE)
## -- Density estimation for bounded data via GMMs -----------
##
## Boundaries: suicide
##        lower       0
##        upper     Inf
##
## Model E (univariate, equal variance) model with 1 component
## on the transformation scale:
##
##  log-likelihood  n df   BIC   ICL
##        -497.82 86  3 -1009 -1009
##
##                             suicide
## Range-power transformation: 0.19293
##
## Mixing probabilities:
## 1
## 1
##
## Means:
##       1
## 6.7001
##
## Variances:
##       1
## 7.7883
```

```
plot(bdens, what = "density")
rug(suicide)              # add data points at the bottom of the graph
abline(v = 0, lty = 3)   # draw a vertical line at the natural boundary
```

The estimated parameter of the range-power transformation is equal to 0.19, and on the transformed scale the optimal GMM is a model with a single mixture component. The corresponding density estimate on the original scale of the variable is shown in Figure 5.9b. As the plot shows, accounting for the natural boundary of the variable better represents the underlying density of the data.

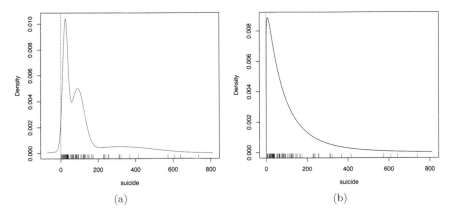

(a)   (b)

FIGURE 5.9: Density estimates for the suicide data. Panel (a) shows the default estimate which ignores the lower boundary of the variable, while panel (b) shows the density estimated accounting for the natural boundary of the variable.

EXAMPLE 5.7: Density estimation of racial data

This dataset provides the proportion of white student enrollment in 56 school districts in Nassau County (Long Island, New York, USA), for the 1992–1993 school year (Simonoff, 1996, Sec. 3.2). Because each observation is a proportion, the density estimate for this data should not fall outside the $[0, 1]$ range.

The following code reads the data, performs density estimation by specifying both the lower and upper bounds of the variable under study, and plots the data and the estimated density:

```
data("racial", package = "mclustAddons")
bdens <- densityMclustBounded(racial$PropWhite, lbound = 0, ubound = 1)
plot(bdens, what = "density",
```

```
    lwd = 2, col = "dodgerblue2",
    data = racial$PropWhite, breaks = 15,
    xlab = "Proportion of white students enrolled in schools")
rug(racial$PropWhite)          # add data points at the bottom of the graph
abline(v = c(0, 1), lty = 3)  # draw a vertical line at the natural boundary
```

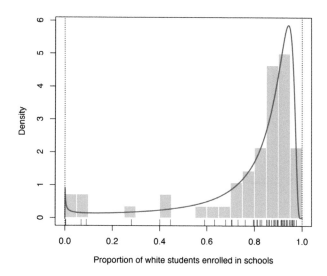

FIGURE 5.10: Density estimate for the racial data obtained by taking into account the natural boundary of the proportion.

Figure 5.10 displays the estimated density, showing that the majority of the schools have at least 70% of white students, but there is also a small peak near the lower boundary containing schools with almost 0% of white students.

5.6   Highest Density Regions

Density curves are a very effective and useful way to show the distribution of values in a dataset. For two-dimensional data, perspective and contour plots represent a standard way to visualize densities (see for instance Figure 5.6). An alternative method is discussed by Hyndman (1996), who proposed the use of Highest Density Regions (HDRs) for summarizing probability distributions. These are defined as the (possibly disjoint) regions of the sample space covering a specified probability level.

Let $f(x)$ be the density function of a random variable X; then the $100(1 -$

$\alpha)\%$ HDR is the subset $R(f_\alpha)$ of the sample space of X such that

$$R(f_\alpha) = \{x : f(x) \geq f_\alpha\},$$

where f_α is the largest value such that $\Pr(X \in R(f_\alpha)) \geq 1 - \alpha$.

For estimating $R(f_\alpha)$, Hyndman (1996) proposed considering the $(1 - \alpha)$-quantile of the density function evaluated at the observed data points. The accuracy of this method depends on the sample size, so for small to moderate sample sizes the degree of accuracy can be improved by enlarging the set of observations with additional simulated data points.

For example, consider the density of a univariate two-component Gaussian mixture defined with the following code and shown in Figure 5.11:

```
f <- function(x)
  0.7*dnorm(x, mean = 0, sd = 1) + 0.3*dnorm(x, mean = 4, sd = 1)
curve(f, from = -4, to = 8)
```

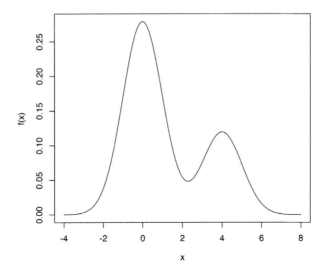

FIGURE 5.11: Density of the univariate two-component Gaussian mixture $f(x) = 0.7\ \phi(x|\mu = 0, \sigma = 1) + 0.3\ \phi(x|\mu = 4, \sigma = 1)$.

A simulated dataset drawn from this mixture can be obtained as follows:

```
par <- list(pro = c(0.7, 0.3), mean = c(0, 4),
            variance = mclustVariance("E", G = 2))
par$variance$sigmasq <- c(1, 1)
x <- sim(modelName = "E", parameters = par, n = 1e4)[, -1]
```

The first column returned by sim() contains the (randomly generated) classification and is therefore omitted. For details of the procedure, see Section 7.4.

The highest density regions corresponding to specified probability levels
can be computed using the function hdrlevels(), and then plotted as follows:

```
prob <- c(0.25, 0.5, 0.75, 0.95)
(hdr <- hdrlevels(f(x), prob))
##        25%        50%        75%        95%
## 0.250682 0.158753 0.096956 0.048063
for (j in seq(prob))
{
   curve(f, from = -4, to = 8)
   mtext(side = 3, paste0(prob[j]*100, "% HDR"), adj = 0)
   abline(h = hdr[j], lty = 2)
   rug(x, col = "lightgrey")
   rug(x[f(x) >= hdr[j]])
}
```

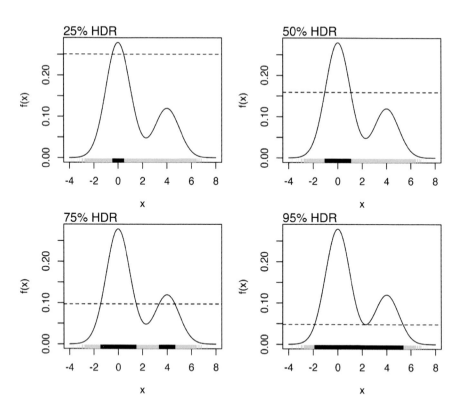

FIGURE 5.12: Highest density regions at specified probability levels from the
density of a univariate two-component Gaussian mixture.

The resulting plot is shown in Figure 5.12.

In general the true density function is unknown, so that an estimate must be used, but the procedure remains essentially unchanged. For instance, we may estimate the density with **mclust** and then obtain the HDR levels as follows:

```
dens <- densityMclust(x, plot = FALSE)
hdrlevels(predict(dens, x), prob)
##       25%       50%       75%       95%
## 0.251234 0.159436 0.096781 0.048798
```

6

Visualizing Gaussian Mixture Models

Graphical displays are used extensively in previous chapters for showing clustering, classification, and density estimation. This chapter further discusses visualization methods for model-based clustering by illustrating fine tuning of plots available in **mclust**. In high-dimensional settings, in addition to marginal projections, several methods are presented for visualizing the clustering or classification structure on a subspace of reduced dimension.

6.1 Displays for Univariate Data

mclust models of univariate data can be visualized with mclust1Dplot().

EXAMPLE 6.1: Additional graphs for the fish length data

Consider the dataset introduced in Example 2.1 on the length (in inches) of 256 snapper fish. The following code fits a four-component varying-variance ("V") GMM to this data with Mclust():

```
data("Snapper", package = "FSAdata")
x <- Snapper[,1]
mod <- Mclust(x, G = 4, modelNames = "V")
summary(mod, parameters = TRUE)
## ----------------------------------------------------
## Gaussian finite mixture model fitted by EM algorithm
## ----------------------------------------------------
##
## Mclust V (univariate, unequal variance) model with 4 components:
##
##   log-likelihood   n df     BIC    ICL
##         -489.28 256 11 -1039.6 -1098
##
## Clustering table:
##    1   2   3   4
##   26 143  56  31
##
```

```
## Mixing probabilities:
##        1        2        3        4
## 0.098263 0.542607 0.177391 0.181739
##
## Means:
##      1      2      3      4
## 3.3631 5.4042 7.5761 8.9021
##
## Variances:
##        1        2        3        4
## 0.073383 0.408370 0.197323 2.839843
```

Classification and uncertainty plots for this model can be produced with the
following code:

```
mclust1Dplot(x, what = "classification",
             parameters = mod$parameters, z = mod$z,
             xlab = "Fish length")
mclust1Dplot(x, what = "uncertainty",
             parameters = mod$parameters, z = mod$z,
             xlab = "Fish length")
```

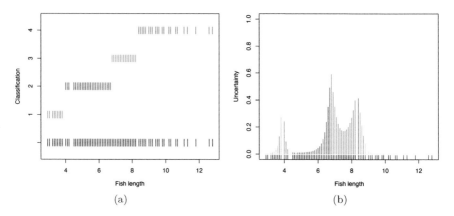

FIGURE 6.1: Classification (a) and uncertainty (b) plots created with
mclust1Dplot() for the GMM fit to the fishery data.

The resulting plots are shown in Figure 6.1. Classification errors and
density for Mclust() models can also be displayed with mclust1Dplot().
Additional arguments can be specified for fine tuning, as described in
help("mclust1Dplot"). MclustDA() and densityMclust() models can also be
vizualized with mclust1Dplot() in the same way as described above.

6.2 Displays for Bivariate Data

mclust models fitted to bivariate data can be visualized with `mclust2Dplot()` and, for finer details, `surfacePlot()`.

EXAMPLE 6.2: Additional graphs for the Old Faithful data

In the following code, classification and uncertainty plots are produced for a `Mclust()` model estimated on the `faithful` dataset discussed in Example 3.4

```
mod <- Mclust(faithful)
mclust2Dplot(data = faithful, what = "classification",
             parameters = mod$parameters, z = mod$z)
mclust2Dplot(data = faithful, what = "uncertainty",
             parameters = mod$parameters, z = mod$z)
```

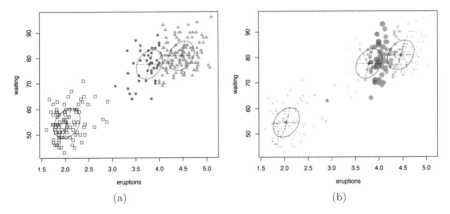

(a)   (b)

FIGURE 6.2: Classification (a) and uncertainty (b) plots created with `mclust2Dplot()` for the model fitted with `Mclust()` to the `faithful` dataset.

The resulting plots are shown in Figure 6.2. In both plots, the ellipses are the multivariate analogs of the standard deviations for each mixture component. In the classification plot, points in different clusters are marked by different symbols and colors. In the uncertainty plot, larger symbols and more opaque shades correspond to higher levels of uncertainty. Data points on the boundaries between the two clusters on the top-right part of the plot are associated with higher values of clustering uncertainty.

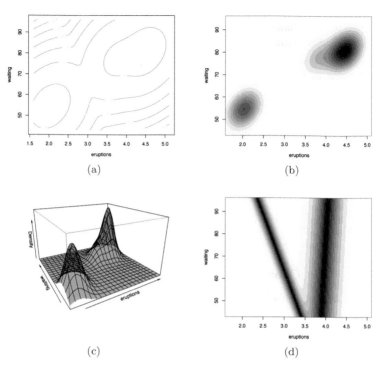

FIGURE 6.3: Density (a–c) and uncertainty (d) surfaces created with `surfacePlot()` for the `faithful` dataset. A logarithmic transformation is used for the density plot in panel (a), which is drawn as a contour surface. No transformation is applied to panels (b) and (c) which show, respectively, an image plot of the density surface and a 3D perspective plot. Panel (d) displays the image surface of the square root of the clustering uncertainty.

Density or uncertainty for **mclust** models of bivariate data can also be displayed using `surfacePlot()`. This function invisibly returns the grid coordinates and the corresponding surface values employed for plotting, information that advanced users can use for further processing. For example, Figure 6.3 shows contour, image, perspective, and uncertainty plots for the model fitted to the `faithful` dataset:

```
surfacePlot(data = faithful, parameters = mod$parameters,
            what = "density", type = "contour",
            transformation = "log")
surfacePlot(data = faithful, parameters = mod$parameters,
            what = "density", type = "image")
surfacePlot(data = faithful, parameters = mod$parameters,
            what = "density", type = "persp")
```

```
surfacePlot(data = faithful, parameters = mod$parameters,
            what = "uncertainty", type = "image",
            transformation = "sqrt")
```

Note that plots in Figure 6.3 are computed over an evenly spaced grid of 200 points (the default) along each axis. More arguments are available for fine tuning; a detailed description can be found via `help("surfacePlot")`.

6.3 Displays for Higher Dimensional Data

6.3.1 Coordinate Projections

Coordinate projections can be plotted in **mclust** with `coordProj()`.

EXAMPLE 6.3: Coordinate projections for the iris data

Consider the best (according to BIC) three-group GMM estimated on the `iris` dataset described in Example 3.13:

```
data("iris", package = "datasets")
mod <- Mclust(iris[,1:4], G = 3)
summary(mod)
## ------------------------------------------------------
## Gaussian finite mixture model fitted by EM algorithm
## ------------------------------------------------------
##
## Mclust VEV (ellipsoidal, equal shape) model with 3 components:
##
##   log-likelihood   n df     BIC     ICL
##          -186.07 150 38 -562.55 -566.47
##
## Clustering table:
##   1  2  3
## 50 45 55
```

The following code produces projection plots for coordinates 2 and 4 of the iris data (see Figure 6.4).

```
coordProj(data = iris[,1:4], dimens = c(2,4), what = "classification",
          parameters = mod$parameters, z = mod$z)
coordProj(data = iris[,1:4], dimens = c(2,4), what = "uncertainty",
          parameters = mod$parameters, z = mod$z)
coordProj(data = iris[,1:4], dimens = c(2,4), what = "error",
```

```
parameters = mod$parameters, z = mod$z, truth = iris$Species)
```

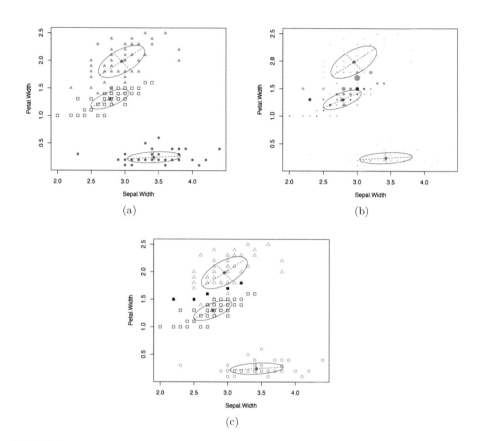

(a)   (b)

(c)

FIGURE 6.4: Coordinate projection plots created with coordProj() for the variables Sepal.Width and Petal.Width from the iris dataset. Panel (a) shows the three-group model-based clustering, with the associated uncertainty in panel (b) and the classification errors in panel (c).

6.3.2   Random Projections

In the previous section, we looked at bivariate marginal coordinate projections that involve selected pairs of variables. Higher dimensional data prove to be much more difficult to visualize. A different approach, discussed in this section, consists of looking at the data from different random perspectives.

Two-dimensional random projections can be computed and plotted in **mclust** with randProj(). These projections are obtained by simulating a set of random orthogonal coordinates, each spanning a two-dimensional projection subspace. **mclust** provides a wrapper function randomOrthogonalMatrix() for generating orthogonal coordinates from a QR factorization of a matrix whose entries are generated randomly from a normal distribution (Heiberger, 1978).

The following code produces a 4×2 orthogonal matrix Q suitable for projecting the iris data:

```
nrow(iris)
## [1] 150
Q <- randomOrthogonalMatrix(ncol(iris[,1:4]), 2)
dim(Q)
## [1] 4 2
QTQ <- crossprod(Q)              # equivalently t(Q) %*% Q
zapsmall(QTQ)                    # 2 x 2 identity matrix
##       [,1] [,2]
## [1,]    1    0
## [2,]    0    1
# projection of iris data onto coordinates Q
irisProj <- as.matrix(iris[,1:4]) %*% Q
```

For a discussion of methods for generating random orthogonal matrices, see Anderson et al. (1987).

The data X can be projected onto a subspace spanned by orthogonal coordinates β by computing $X\beta$. It can be shown that for a given GMM with estimated means $\widehat{\mu}_k$ and covariance matrices $\widehat{\Sigma}_k$ ($k = 1, \ldots, G$), the parameters of the Gaussian components on the projection subspace are $\beta^\top \widehat{\mu}_k$ and $\beta^\top \widehat{\Sigma}_k \beta$, respectively. Finally, note that the number of two-dimensional projections generated by randProj() depends on the number of values provided through the optional argument seeds.

EXAMPLE 6.4: Random projections for the iris data

Consider again the three-group GMM for the iris dataset from the previous subsection. The following code produces four random projections of the iris data with corresponding cluster ellipses and model classification:

```
randProj(data = iris[,1:4], seeds = c(1,13,79,201),
         what = "classification",
         parameters = mod$parameters, z = mod$z)
```

The resulting plots are shown in Figure 6.5. Since the basis of each projection is random, to make the results reproducible we fixed the seeds using the seeds argument in the randProj() function call. If not provided, each call of randProj() will result in a different projection using a system-generated seed.

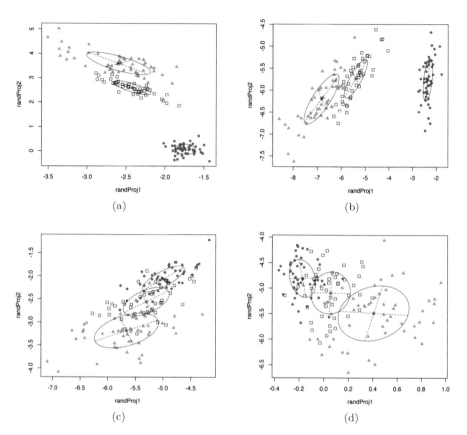

FIGURE 6.5: Random projection plots, all showing the three-group model-based clustering of the iris dataset. Each plot was created with randProj() with a different seed.

6.3.3 Discriminant Coordinate Projections

Both the coordinate and random projection plots decribed above display GMMs in different subspaces, without attempting to select coordinates based on any specific visual criteria. By contrast, *discriminant coordinates* or *crimcoords* (Gnanadesikan, 1977; Flury, 1997) are specifically designed to reveal group separation, and they can be used both in the case of known groups and of groups identified by a clustering model or algorithm.

Consider a data matrix $\boldsymbol{X}$ of dimension $n \times d$, for n observations on d variables, with an associated group structure composed either by G known classes or estimated clusters. Let $\bar{\boldsymbol{x}}$ be the overall sample mean vector, and $\bar{\boldsymbol{x}}_k$ and $\boldsymbol{S}_k$, respectively, the group-specific sample mean vector and sample covariance matrix on n_k observations ($k = 1, \ldots, G$; $n = \sum_{k=1}^{G} n_k$). Then, the

$d \times d$ between-groups covariance matrix

$$\boldsymbol{B} = \frac{1}{G} \sum_{k=1}^{G} n_k (\bar{\boldsymbol{x}}_k - \bar{\boldsymbol{x}})(\bar{\boldsymbol{x}}_k - \bar{\boldsymbol{x}})^\top$$

represents the group separation, while the $d \times d$ pooled within-groups covariance matrix

$$\boldsymbol{W} = \frac{1}{n} \sum_{k=1}^{G} (n_k - 1) \boldsymbol{S}_k.$$

represents the within-group dispersion.

From the point of view of searching for the maximal separation of groups, the optimal projection subspace is given by the set of linear transformations of the original variables that maximizes the ratio of the between-groups covariance to the pooled within-groups covariance. The basis of this projection subspace is spanned by the eigenvectors $\boldsymbol{v}_i$ corresponding to the non-zero eigenvalues of the generalized eigenvalue problem:

$$\boldsymbol{B}\boldsymbol{v}_i = \gamma_i \boldsymbol{W}\boldsymbol{v}_i,$$
$$\boldsymbol{v}_i^\top \boldsymbol{W}\boldsymbol{v}_j = 1 \quad \text{if } i = j, \text{ and 0 otherwise.}$$

The *discriminant coordinates* or *crimcoords* are computed as $\boldsymbol{X}\boldsymbol{V}$, where $\boldsymbol{V} \equiv [\boldsymbol{v}_1, \ldots, \boldsymbol{v}_p]$ is the $d \times p$ matrix of $p = \min(d, G-1), r$ eigenvectors, where r is the number of non-zero eigenvalues. The directions of the discriminant subspace spanned by $\boldsymbol{V}$ are in decreasing order of effectiveness in identifying the separation among groups as expressed by the associated eigenvalues, $\gamma_1 \geq \gamma_2 \geq \cdots \geq \gamma_p > 0$.

Alternatively, unbiased-sample estimates may be used for these compu-tatons: $\frac{G}{G-1}\boldsymbol{B}$ for the between-groups covariance and $\frac{n}{n-G}\boldsymbol{W}$ for the within-groups covariance. These would change the eigenvalues and eigenvectors by a constant of proportionality, equal to $n(G-1)/(G(n-G))$ and $\sqrt{n/(n-G)}$, respectively.

The method described above is implemented in the function `crimcoords()` available in **mclust**. This requires the arguments `data`, a matrix or data frame of observed data, and `classification`, a vector or a factor giving the groups classification, either known class labels or estimated cluster assignments. Optional arguments are `numdir`, an integer value specifying the number of directions to return (by default all those corresponding to non-zero eigenvalues), and `unbiased`, a logical value specifying whether unbiased estimates should be used (by default set to `FALSE` so that MLE estimates are used). `summary()` and `plot()` methods are available to show the estimated projections basis and the corresponding data projection.

EXAMPLE 6.5: Discriminant coordinates for the iris data

Discriminant coordinates of the iris data are computed and plotted with the following code:

```
plot(crimcoords(iris[,1:4], mod$classification))
```

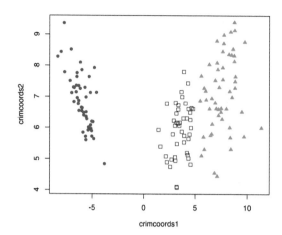

FIGURE 6.6: Discriminant coordinates or crimcoords projection for the clustering for the 3-group model-based clustering of the `iris` dataset.

with the resulting plot shown in Figure 6.6.

EXAMPLE 6.6: Discriminant coordinates for the thyroid disease data

Consider the data introduced in Example 3.2 from the Thyroid Disease Data from the UCI Repository (Dua and Graff, 2017). We compute the discriminant coordinates with `crimcoords()` using the thyroid diagnosis as the classification variable, and then print the computed eigenvectors and eigenvalues using the `summary()` function:

```
data("thyroid", package = "mclust")
CRIMCOORDS <- crimcoords(thyroid[,-1], thyroid$Diagnosis)
summary(CRIMCOORDS)
## -------------------------------------
## Discriminant coordinates (crimcoords)
## -------------------------------------
##
## Estimated basis vectors:
##         crimcoords1 crimcoords2
## RT3U    -0.025187   -0.0019885
## T4       0.307600    0.1042255
## T3       0.116701    0.4371644
## TSH     -0.038588    0.1484593
## DTSH    -0.073398    0.0745668
```

```
##
##              crimcoords1 crimcoords2
## Eigenvalues     275.143      52.473
## Cum. %           83.983     100.000
```

The associated plot() method can then be used for plotting the data points projected onto the discriminant coordinates subspace. In the following code we also add a projection of the corresponding group-centroids:

```
plot(CRIMCOORDS)
points(CRIMCOORDS$means %*% CRIMCOORDS$basis, pch = 3, cex = 1.5, lwd = 2)
legend("topright", legend = levels(thyroid$Diagnosis), inset = 0.02,
       col = mclust.options("classPlotColors")[1:3],
       pch = mclust.options("classPlotSymbols")[1:3])
```

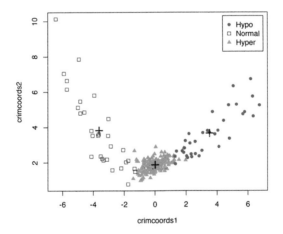

FIGURE 6.7: Discriminant coordinates or crimcoords projection for the diagnosis classes of the thyroid dataset. Group centroids are represented with the + symbol.

The resulting plot is shown in Figure 6.7. It is interesting to note that patients with normal thyroid function line up along the left arm, while the remaining patients with hyperthyroidism or hypothyroidism line up along the right arm.

An alternative method for selecting an informative basis of the projection subspace is discussed in Scrucca and Serafini (2019). They use a projection pursuit approach that maximizes an approximation to the negative entropy for Gaussian mixtures to determine coordinates for display.

6.4   Visualizing Model-Based Clustering and Classification on Projection Subspaces

When the number of clustering or classification features is larger than two, the marginal or random projections described in Section 6.3 are not guaranteed to produce displays that reveal the underlying structure of the data. Discriminant coordinates (Section 6.3.3) provide an attempt at showing maximal separation of groups without explicitly assuming any specific model.

With the aim of visualizing the clustering structure and geometric characteristics induced by a GMM, Scrucca (2010, 2014) proposed a methodology for projecting the data onto subspaces of reduced dimension. These subspaces are spanned by a set of linear combinations of the original variables, called GMMDR directions ("DR" for Dimension Reduction). They are obtained through a procedure that looks for the smallest subspace that captures the clustering information contained in the data. Thus, the goal is to identify those directions where the cluster means $\boldsymbol{\mu}_k$ and/or the cluster covariances $\boldsymbol{\Sigma}_k$ vary as much as possible, provided that each direction is orthogonal to the others in a transformed space. These methods are implemented in the **mclust** functions `MclustDR()` and `MclustDRsubsel()`.

6.4.1   Projection Subspaces for Visualizing Cluster Separation

The variation among the cluster means is captured by the $d \times d$ matrix

$$\boldsymbol{\mathcal{B}} = \sum_{k=1}^{G} \pi_k (\boldsymbol{\mu}_k - \boldsymbol{\mu})(\boldsymbol{\mu}_k - \boldsymbol{\mu})^\top, \tag{6.1}$$

where $\boldsymbol{\mu} = \sum_{k=1}^{G} \pi_k \boldsymbol{\mu}_k$ is the marginal mean vector. The projection subspace associated with this variation is spanned by the eigenvectors $\boldsymbol{v}_i$ corresponding to the non-zero eigenvalues of the following generalized eigenvalue problem:

$$\boldsymbol{\mathcal{K}}_\mu \boldsymbol{v}_i = \gamma_i \boldsymbol{\Sigma} \boldsymbol{v}_i, \tag{6.2}$$

$$\boldsymbol{v}_i^\top \boldsymbol{\Sigma} \boldsymbol{v}_j = 1 \quad \text{if } i = j, \text{ and } 0 \text{ otherwise},$$

in which

$$\boldsymbol{\mathcal{K}}_\mu = \boldsymbol{\mathcal{B}} \boldsymbol{\Sigma}^{-1} \boldsymbol{\mathcal{B}} \tag{6.3}$$

is the kernel matrix associated with the cluster means, and

$$\boldsymbol{\Sigma} = \frac{1}{n} \sum_{i=1}^{n} (\boldsymbol{x}_i - \boldsymbol{\mu})(\boldsymbol{x}_i - \boldsymbol{\mu})^\top \tag{6.4}$$

is the marginal covariance matrix. $\boldsymbol{\Sigma}$ is symmetric positive definite, and $\boldsymbol{\mathcal{K}}_\mu$ is also symmetric with nonnegative eigenvalues, of which $p = \min(d, G - 1)$ are non-zero. The GMMDR variables $X\beta$ are the projection of the $n \times d$ matrix X onto the subspace spanned by $\beta \equiv [v_1, \ldots, v_p]$. The basis vectors v_i are orthogonal in the space transformed by Σ.

EXAMPLE 6.7: Visualizing wine data clustering

In this example, we consider the Italian wines from the package gclus (Hurley, 2019) data previously discussed in Example 3.3, where the estimated VVE model with three components was selected by BIC. We first obtain the Mclust() model for this data and the confusion matrix for the resulting classification:

```
data("wine", package = "gclus")
Class <- factor(wine$Class, levels = 1:3,
                labels = c("Barolo", "Grignolino", "Barbera"))
X <- data.matrix(wine[,-1])
mod <- Mclust(X, G = 3, modelNames = "VVE")
table(Class, Cluster = mod$classification)
##              Cluster
## Class          1  2  3
##    Barolo      59  0  0
##    Grignolino   0 69  2
##    Barbera      0  0 48
```

MclustDR() is then applied to obtain a projection subspace:

```
drmod <- MclustDR(mod, lambda = 1)
summary(drmod)
## -----------------------------------------------------------------
## Dimension reduction for model-based clustering and classification
## -----------------------------------------------------------------
##
## Mixture model type: Mclust (VVE, 3)
##
## Clusters  n
##        1 59
##        2 69
##        3 50
##
## Estimated basis vectors:
##                      Dir1        Dir2
## Alcohol        0.13399009  0.19209123
## Malic         -0.03723778  0.06424412
## Ash           -0.01313103  0.62738796
## Alcalinity    -0.04299147 -0.03715437
## Magnesium     -0.00053971  0.00051772
## Phenols       -0.13507235 -0.04687991
```

```
## Flavonoids         0.51323644 -0.13391186
## Nonflavanoid       0.68462875 -0.61863302
## Proanthocyanins   -0.07506153 -0.04652587
## Intensity         -0.08855450  0.04877118
## Hue                0.28941727 -0.39564601
## OD280              0.36197696 -0.00779361
## Proline            0.00070724  0.00075867
##
##                    Dir1    Dir2
## Eigenvalues       1.6189   1.292
## Cum. %           55.6156 100.000
```

Setting the tuning parameter λ to 1 (the default in MclustDR()) results in a projection subspace for maximal separation of clusters. Other options that incorporate the variation in covariances are discussed in Section 6.4.2.

Recall that the basis from (6.2) consists of $p = \min(d, G - 1)$ directions, where d is the number of variables and G the number of mixture components or clusters. In this example, there are $d = 13$ features and $G = 3$ clusters, so the reduced subspace is two-dimensional. The projected data are shown in Figure 6.8a, obtained with the code:

```
plot(drmod, what = "contour")
```

On the same subspace we can also plot the uncertainty boundaries corresponding to the MAP classification:

```
plot(drmod, what = "boundaries")
```

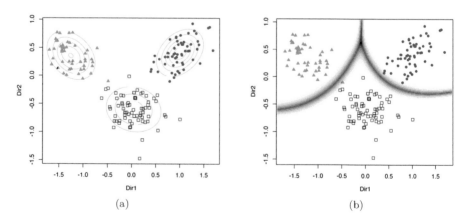

(a)   (b)

FIGURE 6.8: Contour plot of estimated mixture densities (a) and uncertainty boundaries (b) projected onto the subspace estimated with MclustDR() for the wine dataset.

Although the GMMDR subspaces discussed in this section are not guaranteed to show cluster separation, they do provide useful clustering information in many cases of practical interest. When there are only two clusters, there is just one basis vector and the projections will be univariate.

6.4.2 Incorporating Variation in Covariances

Analogous to the scenario for cluster means discussed in the previous section, the variation among the cluster covariances is captured by the $d \times d$ kernel matrix

$$\mathcal{K}_\Sigma = \sum_{k=1}^{G} \pi_k (\mathbf{\Sigma}_k - \bar{\mathbf{\Sigma}}) \mathbf{\Sigma}^{-1} (\mathbf{\Sigma}_k - \bar{\mathbf{\Sigma}})^\top,$$

where

$$\bar{\mathbf{\Sigma}} = \sum_{k=1}^{G} \pi_k \mathbf{\Sigma}_k$$

is the *pooled* within-cluster covariance matrix, and $\mathbf{\Sigma}$ is the marginal covariance matrix in (6.4).

The variation in both means and variances can then be combined, and the associated projection subspace is spanned by the eigenvectors $\mathbf{v}_i$ that solve the generalized eigenvalue problem

$$\left(\lambda \mathcal{K}_\mu + (1 - \lambda) \mathcal{K}_\Sigma \right) \mathbf{v}_i = \gamma_i \mathbf{\Sigma} \mathbf{v}_i, \tag{6.5}$$

$$\mathbf{v}_i^\top \mathbf{\Sigma} \mathbf{v}_j = 1 \quad \text{if } i = j, \text{ and } 0 \text{ otherwise,}$$

in which $\mathcal{K}_\mu$ is the kernel matrix (6.3) associated with the cluster means in Section 6.4.1, and $\lambda \in [0, 1]$ is a parameter controlling the relative contributions of the mean and covariance variations (Scrucca, 2014).

For larger values of λ, the estimated directions will tend to focus on differences in location. For $\lambda = 1$ (the default in `MclustDR()`), differences in class covariances are ignored, and the most discriminant directions (those that show maximal separation among classes) are recovered (Section 6.4.1). For $\lambda = 0.5$, the two types of information are equally weighted (Scrucca, 2010). When $\lambda < 1$, the GMMDR variables $\mathbf{X}\boldsymbol{\beta}$ are the projection of the $n \times d$ matrix $\mathbf{X}$ onto the subspace spanned by $\boldsymbol{\beta} \equiv [\mathbf{v}_1, \ldots, \mathbf{v}_d]$, which has the same dimension as the data, regardless of the number of GMM components. Note that each generalized eigenvalue γ_i is the sum of contributions from both the means and the variances:

$$\gamma_i = \gamma_i \mathbf{v}_i^\top \mathbf{\Sigma} \mathbf{v}_i = \lambda \mathbf{v}_i^\top \mathcal{K}_\mu \mathbf{v}_i + (1 - \lambda) \mathbf{v}_i^\top \mathcal{K}_\Sigma \mathbf{v}_i, \quad i = 1, \ldots, d.$$

EXAMPLE 6.8: Visualizing iris data clustering

As an example of the above methodology, consider again the three-group model for the `iris` dataset described in Section 6.3. Once a GMM model has been

fitted, the subspace estimated by MclustDR() can be used for plots that attempt
to capture most of the clustering structure:

```
mod <- Mclust(iris[, 1:4], G = 3)
drmod <- MclustDR(mod, lambda = .5)
summary(drmod)
## ---------------------------------------------------------------------
## Dimension reduction for model-based clustering and classification
## ---------------------------------------------------------------------
##
## Mixture model type: Mclust (VEV, 3)
##
## Clusters  n
##        1 50
##        2 45
##        3 55
##
## Estimated basis vectors:
##                   Dir1      Dir2      Dir3      Dir4
## Sepal.Length  0.14546 -0.220270  0.65191 -0.43651
## Sepal.Width   0.52097  0.097857  0.25527  0.57873
## Petal.Length -0.62095 -0.293850 -0.44704  0.46137
## Petal.Width  -0.56732  0.924963  0.55678 -0.51153
##
##                  Dir1     Dir2      Dir3       Dir4
## Eigenvalues  0.94861  0.62354  0.074295   0.032765
## Cum. %      56.49126 93.62436 98.048774 100.000000
```

The basis vectors spanning the reduced subspace are available through the
summary() method for MclustDR(). These basis vectors are expressed as linear
combinations of the original features, ordered by importance via the associated
generalized eigenvalues (6.5). In this case the first two directions account for
most of the clustering structure. A plot of the eigenvalues, shown in Figure 6.9,
is obtained with the following code:

```
plot(drmod, what = "evalues")
```

According to the previous discussion on the contributions to the generalized
eigenvalues, we would expect to see a separation among the groups along
the first two directions only, with the first associated with differences in
location, and the second associated with differences in spread. This is confirmed
by a scatterplot matrix of data projected onto the estimated subspace (see
Figure 6.10) obtained as follows:

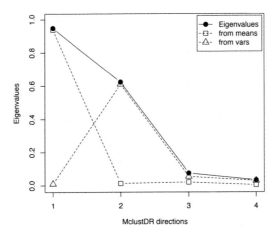

FIGURE 6.9: Plot of generalized eigenvalues from `MclustDR()` and the corresponding contributions from means and variances for the 3-group model-based clustering of the `iris` dataset.

```
plot(drmod, what = "pairs")
```

As already mentioned, the last two directions appear negligible in this example. This may be further investigated by applying the greedy subset selection step proposed in Scrucca (2010, Section 3) that applies the subset selection method of Raftery and Dean (2006) to prune the subset of GMMDR features:

```
sdrmod <- MclustDRsubsel(drmod, verbose = TRUE)
##
## Cycle 1 ...
##
##    Variable Model G     BIC BIC.dif
## 1     Dir1    E 3 -374.60 187.035
## 2     Dir2  EEI 4 -315.66  45.242
## 3     Dir3  VVE 3 -395.01  37.666
##
## Cycle 2 ...
##
##    Variable Model G     BIC BIC.dif
## 1     Dir1    E 3 -354.64 186.167
## 2     Dir2  VVI 3 -294.07  47.343
## 3     Dir3  VVE 3 -389.47  21.632
summary(sdrmod)
## -------------------------------------------------------------
```

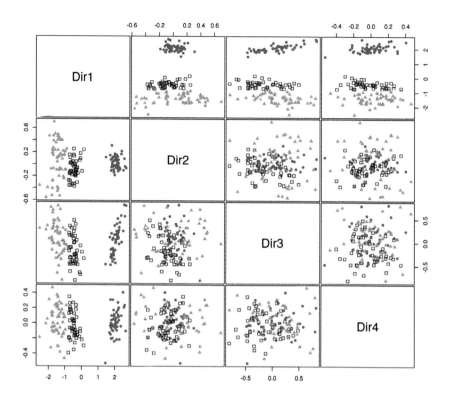

FIGURE 6.10: The iris data projected onto the principal eigenvectors from MclustDR(). The colors and symbols correspond to the 3-group model-based clustering.

```
## Dimension reduction for model-based clustering and classification
## ------------------------------------------------------------------
##
## Mixture model type: Mclust (VVE, 3)
##
## Clusters  n
##        1 50
##        2 49
##        3 51
##
## Estimated basis vectors:
##                  Dir1     Dir2     Dir3
## Sepal.Length  0.10835 -0.18789  0.71035
## Sepal.Width   0.54164  0.10527  0.25902
```

```
## Petal.Length -0.65930 -0.30454 -0.43979
## Petal.Width  -0.51010  0.92783  0.48465
##
##                   Dir1     Dir2      Dir3
## Eigenvalues  0.94257  0.64868   0.03395
## Cum. %       57.99696 97.91101 100.00000
```

The subset selection chooses three directions and essentially the first three previously obtained:

```
zapsmall(cor(drmod$dir, sdrmod$dir)^2)
##          Dir1    Dir2    Dir3
## Dir1 0.99975 0.00020 0.00006
## Dir2 0.00021 0.99537 0.00443
## Dir3 0.00004 0.00444 0.99552
## Dir4 0.00000 0.00000 0.00000
```

In the following plots, we use the default argument dimens = c(1, 2) to project onto the first two GMMDR directions, which are of most interest for visualizing the clustering structure:

```
plot(sdrmod, what = "contour", nlevels = 7)
plot(sdrmod, what = "classification")
plot(sdrmod, what = "boundaries")
```

The first command draws a bivariate contour plot of the mixture component densities, the second command draws the classification boundaries based on the MAP principle, while the last command draws a scatterplot showing uncertainty boundaries (panels (a), (b), and (c) in Figure 6.11).

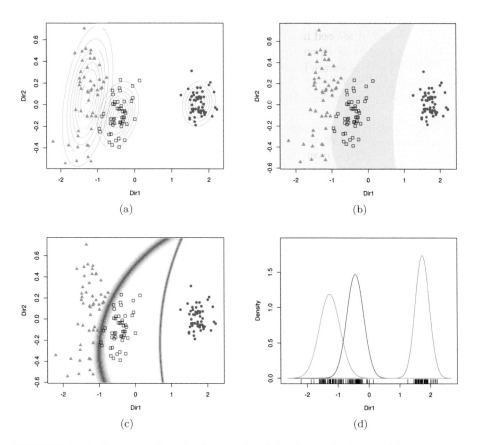

FIGURE 6.11: Contour plot of mixture densities for each cluster (a), classification regions (b), and uncertainty boundaries (c) drawn on the first two GMMDR directions of the projection subspace estimated with `MclustDRsubsel()` for the 3-group model-based clustering of the `iris` dataset. Panel (d) shows the conditional densities along the first GMMDR direction.

We can also produce a plot of the densities conditional on the estimated cluster membership for the first GMMDR direction as follows (panel (d) in Figure 6.11):

```
plot(sdrmod, what = "density")
```

Other plots can be obtained via the argument `what`, and fine tuning for some parameters is also available; see `help("plot.MclustDR")` for a comprehensive list and examples.

6.4.3 Projection Subspaces for Classification

The approach discussed in the previous sections for clustering has been further extended to the case of supervised classification (Scrucca, 2014). In this scenario, the fact that classes are known and that they can be made up of one or more mixture components must be taken into account.

EXAMPLE 6.9: Visualizing Swiss banknote classification

The banknote dataset containing six physical measurements of a sample of Swiss Franc bills is given in Tables 1.1 and 1.2 of Flury and Riedwyl (1988). One hundred banknotes were classified as genuine and 100 as counterfeits.

We fit a MclustDA() classification model to this data:

```
data("banknote", package = "mclust")
mod <- MclustDA(data = banknote[, -1], class = banknote$Status)
summary(mod)
## ------------------------------------------------
## Gaussian finite mixture model for classification
## ------------------------------------------------
##
## MclustDA model summary:
##
##   log-likelihood   n df     BIC
##          -646.08 200 66 -1641.8
##
## Classes          n  % Model G
##   counterfeit 100 50   EVE 2
##   genuine     100 50   XXX 1
##
## Training confusion matrix:
##             Predicted
## Class        counterfeit genuine
##   counterfeit         100       0
##   genuine               0     100
## Classification error = 0
## Brier score          = 0
```

We then apply MclustDR() to obtain a projection subspace:

```
drmod <- MclustDR(mod, lambda = .5)
summary(drmod)
## --------------------------------------------------------------------
## Dimension reduction for model-based clustering and classification
## --------------------------------------------------------------------
##
```

```
## Mixture model type: MclustDA
##
## Classes         n Model G
##   counterfeit 100   EVE 2
##   genuine     100   XXX 1
##
## Estimated basis vectors:
##                Dir1      Dir2      Dir3      Dir4       Dir5      Dir6
## Length    -0.10139 -0.328225  0.797068 -0.033629 -0.3174275  0.085062
## Left      -0.21718 -0.305014 -0.303111 -0.893349  0.3700659 -0.565410
## Right      0.29222 -0.018401 -0.495891  0.407413 -0.8612986  0.480799
## Bottom     0.57591  0.445352  0.120173 -0.034595  0.0043174 -0.078640
## Top        0.57542  0.385535  0.100865 -0.103623  0.1359128  0.625902
## Diagonal  -0.44089  0.672250 -0.047784 -0.151252 -0.0443255  0.209691
##
##                Dir1      Dir2      Dir3      Dir4       Dir5       Dir6
## Eigenvalues 0.87242   0.55373   0.48546   0.13291   0.053075   0.027273
## Cum. %     41.05755 67.11689  89.96377  96.21866  98.716489 100.000000
```

A plot of the generalized eigenvalues associated with the estimated directions
is obtained using:

```
plot(drmod, what = "evalues")
```

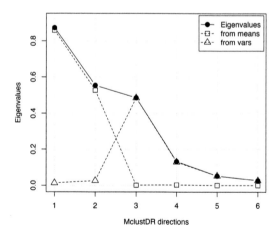

FIGURE 6.12: Plot of generalized eigenvalues from MclustDR() applied to a
classification model for the banknote data.

This suggests that the first two directions mainly contain information on class
separation, with the remaining dimensions showing differences in variances.

  A pairs plot of the data points projected along these directions seems to suggest that a further reduction could be achieved by selecting a subset of them (see Figure 6.13):

```
plot(drmod, what = "pairs", lower.panel = NULL)
clPairsLegend(0.1, 0.4, class = levels(drmod$classification),
              col = mclust.options("classPlotColors")[1:2],
              pch = mclust.options("classPlotSymbols")[1:2],
              title = "Swiss banknote data")
```

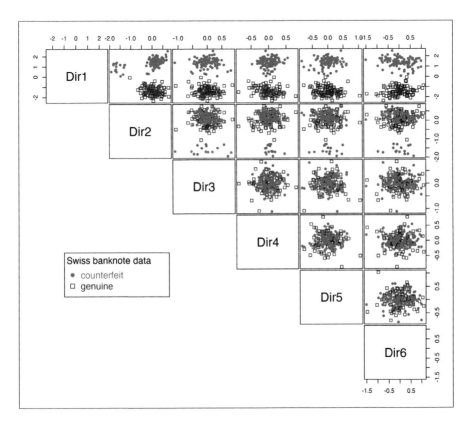

FIGURE 6.13: Pairs plot of points projected onto the directions estimated with `MclustDR()` for the banknote dataset.

This option can be investigated by applying the subset selection procedure described in Scrucca (2010) with the code:

```
sdrmod <- MclustDRsubsel(drmod, verbose = TRUE)
##
## Cycle 1 ...
##
```

```
##   Variable   Model   G     BIC BIC.dif
## 1    Dir1     E|X 2|1 -521.81  212.03
## 2    Dir2 EEE|XXX 2|1 -662.39  192.65
##
## Cycle 2 ...
##
##   Variable   Model   G     BIC BIC.dif
## 1    Dir1     E|X 2|1 -523.63  210.17
## 2    Dir2 EEE|XXX 2|1 -662.36  194.51
summary(sdrmod)
## ----------------------------------------------------------------
## Dimension reduction for model-based clustering and classification
## ----------------------------------------------------------------
##
## Mixture model type: MclustDA
##
## Classes          n Model G
##    counterfeit 100   EEE 2
##    genuine     100   XXX 1
##
## Estimated basis vectors:
##               Dir1     Dir2
## Length   -0.10552 -0.328204
## Left     -0.22054 -0.304765
## Right     0.29083 -0.018957
## Bottom    0.57981  0.444505
## Top       0.57850  0.384658
## Diagonal -0.42989  0.673420
##
##                 Dir1      Dir2
## Eigenvalues  0.85929   0.53687
## Cum. %      61.54666 100.00000
```

The subset selection procedure identifies only two directions as being important, essentially the first two directions previously obtained:

```
zapsmall(cor(drmod$dir, sdrmod$dir)^2)
##          Dir1    Dir2
## Dir1 0.99997 0.00003
## Dir2 0.00003 0.99997
## Dir3 0.00000 0.00000
## Dir4 0.00000 0.00000
## Dir5 0.00000 0.00000
## Dir6 0.00000 0.00000
```

Summary plots can then be obtained on the 2-dimensional estimated subspace as follows:

```
plot(sdrmod, what = "contour", nlevels = 15)
plot(sdrmod, what = "classification")
```

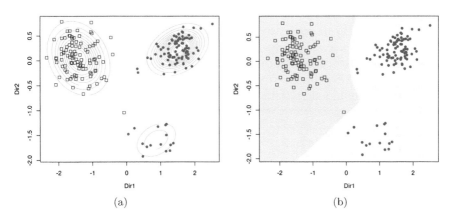

(a)　　　　　　　　　　　　　　(b)

FIGURE 6.14: Contour plot of mixture densities for each class (a) and MAP classification regions (b) drawn on the projection subspace estimated with MclustDRsubsel() for the banknote dataset.

These plots show, respectively, a contour plot of the mixture component densities for each class, and the classification regions based on the MAP principle (see Figure 6.14). The group of counterfeit banknotes is clearly composed of two distinct subgroups, whereas the genuine banknotes appear as a homogeneous group with the presence of an outlying note.

6.4.4　Relationship to Other Methods

Approaches analogous to projection subspaces have been proposed for the cases of finite mixtures of multivariate t distributions (Morris et al., 2013), mixtures of shifted asymmetric Laplace distributions (Morris and McNicholas, 2013), and generalized hyperbolic mixtures (Morris and McNicholas, 2016).

The MclustDR() directions, obtained using the generalized eigendecompositions discussed in Section 6.4.1 and in Section 6.4.2 for the specific case $\lambda = 1$ in (6.5), are essentially equivalent to the discriminant coordinates or crimcoords described in Section 6.3.3 for GMMs with one mixture component per group. Within **mclust**, this applies to Mclust() clustering models, EDDA classification models, and MclustDA classification models with one component (but possibly different covariance structures) per class.

6.5   Using ggplot2 with mclust

All of the plots produced by **mclust** and discussed so far use the base plotting system in R. There are historical reasons for this, but it also helps achieve one of the main goals of the package, namely to provide simple, fast, accurate, and nice-looking plots without introducing further dependences on other packages.

ggplot2 is a popular R package for data visualization based on the *Grammar of Graphics* (Wilkinson, 2005). The **ggplot2** package facilitates creation of production-quality statistical graphics. For a comprehensive introduction, see Wickham (2016). Experienced R users can easily produce **ggplot2** plots using the information contained in the objects returned by **mclust** functions, by collecting them in a data frame and then (gg)plotting them using a suitable plot

EXAMPLE 6.10: ggplot2 graphs of Old Faithful data clustering

As a first example, consider the following code which produces the usual classification plot (see Figure 6.15):

```
mod <- Mclust(faithful)
DF <- data.frame(mod$data, cluster = factor(mod$classification))
library("ggplot2")
ggplot(DF, aes(x = eruptions, y = waiting,
                colour = cluster, shape = cluster)) +
  geom_point()
```

A more complex example involves plotting the table of BIC values (see Figure 6.16). In this case, we need to convert the object into a data.frame (or a tibble in *tidyverse* — see Wickham et al. (2019)) by reshaping it from the "wide" to the "long" format. The latter step can be accomplished in several ways; here, we use the **tidyr** R package (Wickham and Henry, 2022):

```
library("tidyr")
DF <- data.frame(mod$BIC[], G = 1:nrow(mod$BIC))
DF <- pivot_longer(DF, cols = 1:14, names_to = "Model", values_to = "BIC")
DF$Model <- factor(DF$Model, levels = mclust.options("emModelNames"))
ggplot(DF, aes(x = G, y = BIC, colour = Model, shape = Model)) +
  geom_point() +
  geom_line() +
  scale_shape_manual(values = mclust.options("bicPlotSymbols")) +
  scale_color_manual(values = mclust.options("bicPlotColors")) +
  scale_x_continuous(breaks = unique(DF$G)) +
  xlab("Number of mixture components") +
  guides(shape = guide_legend(ncol=2))
```

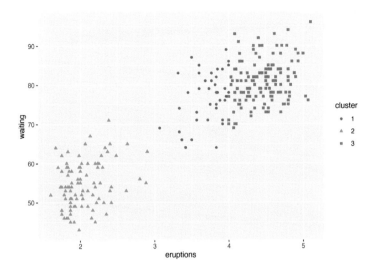

FIGURE 6.15: Scatterplot of the `faithful` data with points marked according to the GMM clusters identified by `Mclust`.

As another example, we may draw a *latent profiles plot* of estimated means for each variable by cluster membership. Consider the following clustering model for the `iris` data:

```
mod <- Mclust(iris[, 1:4], G = 3)
```

A latent profiles plot can be obtained by extracting the component means, reshaping it in a "long" format, and then drawing the desired plot shown in Figure 6.17:

```
means <- data.frame(Profile = 1:mod$G, t(mod$parameters$mean))
means <- pivot_longer(means, cols = -1,
                      names_to = "Variable",
                      values_to = "Mean")
means$Profile  <- factor(means$Profile)
means$Variable <- factor(means$Variable,
                         levels = rownames(mod$parameters$mean))
means
## # A tibble: 12 x 3
##     Profile Variable       Mean
##     <fct>   <fct>         <dbl>
## 1 1         Sepal.Length 5.01
## 2 1         Sepal.Width  3.43
## 3 1         Petal.Length 1.46
## 4 1         Petal.Width  0.246
```

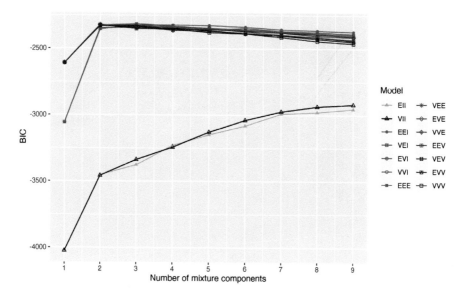

FIGURE 6.16: BIC traces for the GMMs estimated for the faithful data.

```
##  5 2         Sepal.Length 5.92
##  6 2         Sepal.Width  2.78
##  7 2         Petal.Length 4.20
##  8 2         Petal.Width  1.30
##  9 3         Sepal.Length 6.55
## 10 3         Sepal.Width  2.95
## 11 3         Petal.Length 5.48
## 12 3         Petal.Width  1.99
ggplot(means, aes(Variable, Mean, group = Profile,
                  shape = Profile, color = Profile)) +
  geom_point(size = 2) +
  geom_line() +
  labs(x = NULL, y = "Latent profiles means") +
  scale_color_manual(values = mclust.options("classPlotColors")) +
  theme(axis.text.x = element_text(angle = 45, hjust = 1),
        legend.position = "bottom")
```

Note that for a plot like the one in Figure 6.17 to make sense, all variables must be expressed in the same unit of measurement (as in the example above), or else they must be scaled to a common unit, for instance by a preliminary standardization. Ordering of variables along the x-axis is arbitrary, so a user must carefully choose the most appropriate order for the data under study.

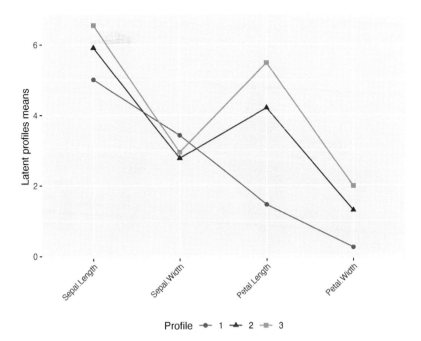

FIGURE 6.17: Latent profiles plot for the (VEV,3) model estimated for the iris data.

EXAMPLE 6.11: ggplot2 graphs of iris data classification

The plotting facilities of **ggplot2** can be used with any **mclust** model. For instance, the following code estimates a classification model for the iris data, extracts the first two GMMDR directions using the methodology described in Section 6.4, and then produces a two-dimensional scatterplot with added convex hulls for the various iris classes:

```
damod <- MclustDA(iris[, 1:4], iris$Species)
drmod <- MclustDR(damod)
DF1 <- data.frame(drmod$dir[, 1:2], class = damod$class)
DF2 <- do.call("rbind", by(DF1, DF1[, 3],
                           function(x) x[chull(x), ]))
ggplot() +
  geom_point(data = DF1,
             aes(x = Dir1, y = Dir2, color = class, shape = class)) +
  geom_polygon(data = DF2,
               aes(x = Dir1, y = Dir2, fill = class),
               alpha = 0.3) +
  scale_color_manual(values = mclust.options("classPlotColors")) +
```

```
scale_fill_manual(values = mclust.options("classPlotColors")) +
scale_shape_manual(values = mclust.options("classPlotSymbols"))
```

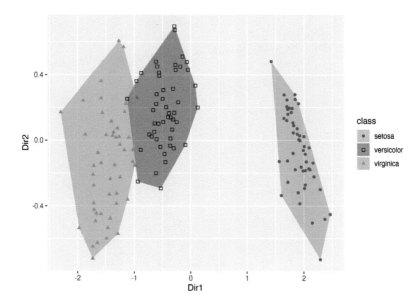

FIGURE 6.18: Scatterplot of the first two GMMDR directions with added convex hulls for the `iris` data classes.

The resulting plot is shown in Figure 6.18. Notice that in the code above we created two different data frames, one used for plotting points and one containing the vertices of the convex hull, the latter computed using the `chull()` function in base R.

EXAMPLE 6.12: ggplot2 graph of density estimate of the waiting time from the Old Faithful data

Now consider the case of plotting a univariate density estimate obtained via `densityMclust()`. The following code draws a histogram of waiting times from the `faithful` data frame, then adds the estimated density. The latter is obtained by first creating a data frame x of equispaced grid points and the corresponding densities computed using the `predict` method associated with the `densityMclust` object. The resulting data frame is then used in `geom_line()` to draw the density estimate.

```
mod <- densityMclust(faithful$waiting, plot = FALSE)
x <- extendrange(faithful$waiting, f = 0.1)
x <- seq(x[1], x[2], length.out = 101)
pred <- data.frame(x, density = predict(mod, newdata = x))
```

```
ggplot(faithful, aes(waiting)) +
  geom_histogram(aes(y = stat(density)), bins = 15,
                 fill = "slategray3", colour = "grey92") +
  geom_line(data = pred, aes(x, density))
```

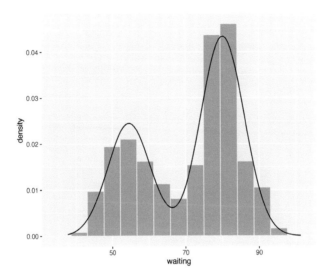

FIGURE 6.19: Plot of histogram and density estimated by densityMclust() for the waiting time of the faithful data.

EXAMPLE 6.13: ggplot2 graphs of bootstrap distributions for the hemophilia data

A final example involves "faceting", an efficient technique for presenting information in panels conditioning on one or more variables. Consider the bootstrap procedure discussed in Example 3.7, where the MclustBootstrap() function is applied to the two-component VVV model for the hemophilia data from the **rrcov** package (Todorov, 2022). Using the information returned and stored in boot, a **ggplot2** plot of the bootstrap distribution for the mixing proportions can be obtained as follows:

```
data("hemophilia", package = "rrcov")
X <- hemophilia[, 1:2]
mod <- Mclust(X, G = 2, modelName = "VVV")
boot <- MclustBootstrap(mod, nboot = 999, type = "bs")
DF <- data.frame(mixcomp = rep(1:boot$G, each = boot$nboot),
                 pro = as.vector(boot$pro))
```

```
ggplot(DF, aes(x = pro)) +
  geom_histogram(aes(y = stat(density)), bins = 15,
                 fill = "slategray3", colour = "grey92") +
  facet_grid(~ mixcomp) +
  xlab("Mixing proportions") +
  ylab("Density of bootstrap distribution")
```

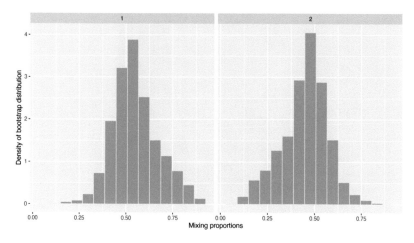

FIGURE 6.20: Bootstrap distribution for the mixture proportions of (VVV,2) model fitted to the `hemophilia` data.

The resulting graph is shown in Figure 6.20. Note that the code above uses `facet_grid()` to obtain a panel for each of the bootstrap distributions.

The same function can also be used for conditioning on more than one variable. For instance, the following code produces Figure 6.21 showing the bootstrap distribution conditioning on both the mixture components (along the rows) and the variables (along the columns):

```
DF <- rbind(
  data.frame("mixcomp"  = 1,
             "variable" = rep(colnames(boot$mean[, , 1]),
                              each = dim(boot$mean)[1]),
             "mean"     = as.vector(boot$mean[, , 1])),
  data.frame("mixcomp"  = 2,
             "variable" = rep(colnames(boot$mean[, , 2]),
                              each = dim(boot$mean)[1]),
             "mean"     = as.vector(boot$mean[, , 2])))
ggplot(DF, aes(x = mean)) +
  geom_histogram(aes(y = stat(density)), bins = 15,
                 fill = "slategray3", colour = "grey92") +
```

```
facet_grid(mixcomp ~ variable, scales = "free_x") +
xlab("Means of mixture") +
ylab("Density of bootstrap distribution")
```

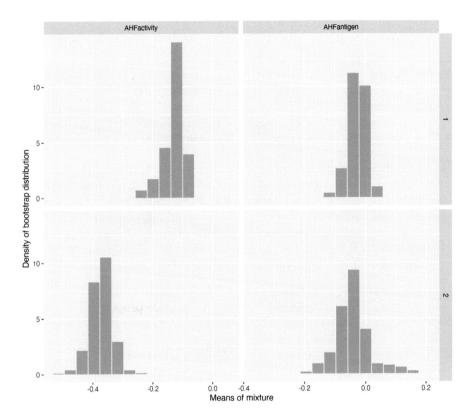

FIGURE 6.21: Bootstrap distribution for the mixture component means of (VVV,2) model fitted to the hemophilia data.

As there are a great many possible options for customizing plots (such as themes and aesthetics), each user may want to experiment to get a suitable plot.

6.6 Using Color-Blind-Friendly Palettes

Most of the plots produced by **mclust** use colors that by default are defined by the following options:

```
mclust.options("bicPlotColors")
##        EII       VII       EEI       EVI       VEI       VVI       EEE
##     "gray"   "black" "#218B21" "#41884F" "#508476" "#58819C" "#597DC3"
##        VEE       EVE       VVE       EEV       VEV       EVV       VVV
## "#5178EA" "#716EE7" "#9B60B8" "#B2508B" "#C03F60" "#C82A36" "#CC0000"
##          E         V
##     "gray"   "black"
mclust.options("classPlotColors")
##  [1] "dodgerblue2"   "red3"            "green3"          "slateblue"
##  [5] "darkorange"    "skyblue1"        "violetred4"      "forestgreen"
##  [9] "steelblue4"    "slategrey"       "brown"           "black"
## [13] "darkseagreen"  "darkgoldenrod3"  "olivedrab"       "royalblue"
## [17] "tomato4"       "cyan2"           "springgreen2"
```

The first option controls the colors to be used for plotting the BIC, ICL, and similar curves, whereas the second option is used to assign colors associated with clusters or classes when plotting data. These colors have been chosen to facilitate visual interpretation of the plots. However, for color-blind individuals, plots with these default colors may be problematic.

Starting with R version 4.0, the function palette.colors() can be used for retrieving colors from some pre-defined palettes. For instance

```
palette.colors(palette = "Okabe-Ito")
##        black       orange      skyblue   bluishgreen       yellow
##    "#000000"    "#E69F00"    "#56B4E9"    "#009E73"    "#F0E442"
##         blue    vermillion reddishpurple         gray
##    "#0072B2"    "#D55E00"    "#CC79A7"    "#999999"
```

returns a color-blind-friendly palette proposed by Okabe and Ito (2008) for individuals suffering from protanopia or deuteranopia, the two most common forms of inherited color blindness; see also Wong (2011).

EXAMPLE 6.14: Using color-blind-friendly palette in mclust for the iris data

A palette suitable for color vision deficiencies can thus be defined and used as the default in **mclust** with the following code:

```
# get and save default palettes
bicPlotColors <- mclust.options("bicPlotColors")
classPlotColors <- mclust.options("classPlotColors")
# set Okabe-Ito palette for use in mclust
bicPlotColors_Okabe_Ito <-
    palette.colors(palette = "Okabe-Ito")[c(9, 1, 2:8, 2:6, 9, 1)]
names(bicPlotColors_Okabe_Ito) <- names(bicPlotColors)
classPlotColorsWong <- palette.colors(palette = "Okabe-Ito")[-1]
```

```
mclust.options("bicPlotColors" = bicPlotColors_Okabe_Ito)
mclust.options("classPlotColors" = classPlotColorsWong)
```

All of the plots subsequently produced by **mclust** functions will use this palette. For instance, the following code produces the plots in Figure 6.22.

```
mod <- Mclust(iris[,3:4])
plot(mod, what = "BIC")
plot(mod, what = "classification")
```

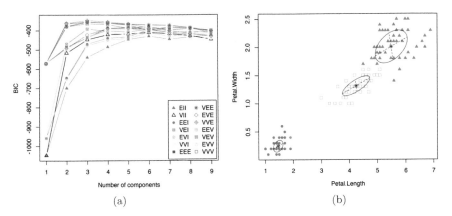

FIGURE 6.22: **mclust** plots with a color-blind-friendly palette.

To restore the default **mclust** palettes use:

```
mclust.options("bicPlotColors" = bicPlotColors)
mclust.options("classPlotColors" = classPlotColors)
```

For more advanced treatment of color issues, the package **colorspace** (Zeileis et al., 2020; Ihaka et al., 2022) is available for manipulating and assessing colors and palettes in R.

7

Miscellanea

In this chapter a range of other issues is discussed, including accounting for outliers and noise. The use of Bayesian methods is presented for avoiding singularities in mixture modeling by adding a prior. The problem of non-Gaussian clusters is addressed by introducing two approaches, one based on combining Gaussian mixture components according to an entropy criterion and one based on identifying connected components. Simulation from mixture models is also discussed briefly, as well as handling of large datasets, high-dimensional data, and missing data. Several data examples are used for illustrating the implementation of the methods in **mclust**.

7.1 Accounting for Noise and Outliers

In the finite mixture framework, noisy data are characterized by the presence of outlying observations that do not belong to any mixture component. There are two main ways to accommodate noise and outliers in a mixture model:

- Adding one or more components to the mixture to represent noise and modifying the EM algorithm accordingly to estimate parameters (Dasgupta and Raftery, 1998; Fraley and Raftery, 1998).

- Modeling by mixtures of distributions with heavier tails than the Gaussian, such as a mixture of t distributions (McLachlan and Peel, 2000).

Some other alternatives for robust Gaussian mixture modeling are proposed in García-Escudero et al. (2008), Punzo and McNicholas (2016), Coretto and Hennig (2016), and Dotto and Farcomeni (2019).

In **mclust**, the strategy for accommodating noise is to include a constant–rate Poisson process mixture component to represent the noise, resulting in the following mixture log-likelihood

$$\ell(\boldsymbol{\Psi}, \pi_0) = \sum_{i=1}^{n} \log \left\{ \frac{\pi_0}{V} + \sum_{k=1}^{G} \pi_k \phi(\boldsymbol{x}_i \mid \boldsymbol{\mu}_k, \boldsymbol{\Sigma}_k) \right\}, \tag{7.1}$$

where V is the hypervolume of the data region, and $\pi_k \geq 0$ are the mixing weights under the constraint $\sum_{k=0}^{G} \pi_k = 1$. An observation contributes $1/V$ to the likelihood if it belongs to the noise component; otherwise its contribution comes from the Gaussian components. This model has been used successfully in a number of applications (Banfield and Raftery, 1993; Dasgupta and Raftery, 1998; Campbell et al., 1997, 1999).

The model-fitting procedure is as follows. Given a preliminary noise assignment for the observations, model-based hierarchical clustering is applied to the correspondingly denoised data to obtain initial clustering partitions for the Gaussian portion of the mixture model. EM is then initialized, using the preliminary noise, together with the hierarchical clustering assignment for each number of clusters specified.

The effectiveness of this approach hinges on obtaining a good initial specification of the noise. Some possible strategies for initial denoising include methods based on Voronoï tessellation (Allard and Fraley, 1997), nearest neighbors (Byers and Raftery, 1998), and robust covariance estimation (Wang and Raftery, 2002).

EXAMPLE 7.1: Clustering with noisy minefields data

Dasgupta and Raftery (1998) discussed the problem of detecting surface-laid minefields from an aircraft image. The main goal was to distinguish the mines from the clutter (metal objects or rocks). As an example, consider the simulated minefield data with clutter analyzed in Fraley and Raftery (1998) and available in **mclust**.

```
data("chevron", package = "mclust")
summary(chevron)
##     class          x                    y
##  data :350    Min.   :  1.18   Min.   :  1.01
##  noise:754    1st Qu.: 34.39   1st Qu.: 27.51
##               Median : 58.82   Median : 49.50
##               Mean   : 61.09   Mean   : 56.71
##               3rd Qu.: 86.52   3rd Qu.: 86.67
##               Max.   :127.49   Max.   :127.86
noise <- with(chevron, class == "noise")
X <- chevron[,2:3]
plot(X, cex = 0.5)
plot(X, cex = 0.5, col = ifelse(noise, "grey", "black"),
     pch = ifelse(noise, 3, 1))
```

The plot of simulated minefield data is shown in panel (a) of Figure 7.1, while panel (b) displays the dataset with noise (clutter) marked by a grey cross. Note that about 70% of the data points are clutter, and the true minefield is contained within the chevron-shaped area, consisting of two intersecting rectangles.

To include noise in modeling with Mclust() or mclustBIC(), an initial guess

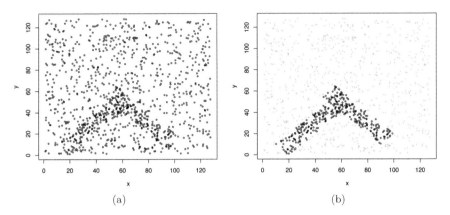

(a)   (b)

FIGURE 7.1: Simulated minefield data (a), marked to distinguish the known noise (b).

of the noise observations must be supplied via the `noise` component of the `initialization` argument . The function `NNclean()` in the R package **prabclus** (Hennig and Hausdorf, 2020) is an implementation of the methodology proposed by Byers and Raftery (1998). Under the assumption that the noise is distributed as a homogeneous Poisson point process and that the remaining data is also distributed as a homogeneous Poisson process but with higher intensity on a (possibly disconnected) subregion, the observed kth nearest neighbor distances are modeled as a mixture of two transformed Gamma distributions with parameters estimated by the EM algorithm. The observed points can then be assigned to either the noise or higher-density component depending on the estimated posterior probabilities.

```
library("prabclus")
nnc <- NNclean(X, k = 5)
table(nnc$z)
##
##   0   1
## 662 442
clPairs(X, nnc$z, colors = c("darkgrey", "black"), symbols = c(3, 1))
```

This example uses $k = 5$ nearest neighbors to detect the noise, which amounts to $662/1104 \approx 60\%$ of the observed data points. The resulting denoised data is shown in Figure 7.2a. With this initial estimate of the noise, a GMM is fitted with the following code:

```
modNoise <- Mclust(X, initialization = list(noise = (nnc$z == 0)))
summary(modNoise$BIC)
```

```
## Best BIC values:
##              EEV,2        EVE,2        EVV,2
## BIC       -21128 -21128.2301 -21134.7385
## BIC diff       0     -0.4463      -6.9546
```

The model-based clustering procedure with the noise component selects the (EEV,2) model, closely followed by the (EVE,2) model. In both cases two clusters are correctly identified. A summary of the estimated model with the highest BIC and the corresponding classification plot are obtained as follows:

```
summary(modNoise, parameters = TRUE)
## ------------------------------------------------------
## Gaussian finite mixture model fitted by EM algorithm
## ------------------------------------------------------
##
## Mclust EEV (ellipsoidal, equal volume and shape) model with 2
## components and a noise term:
##
##   log-likelihood    n df   BIC    ICL
##           -10525 1104 11 -21128 -21507
##
## Clustering table:
##   1   2   0
## 160 151 793
##
## Mixing probabilities:
##        1        2        0
## 0.12290 0.12618 0.75091
##
## Means:
##      [,1]   [,2]
## x 71.473 39.519
## y 36.310 31.566
##
## Variances:
## [,,1]
##         x       y
## x  195.85 -176.99
## y -176.99  197.30
## [,,2]
##         x      y
## x 183.00 176.47
## y 176.47 210.15
##
## Hypervolume of noise component:
```

```
## 16022
addmargins(table(chevron$class, modNoise$classification), 2)
##
##               0   1   2 Sum
##    data      50 156 144 350
##    noise    743   4   7 754
plot(modNoise, what = "classification")
```

The summary output consists of the usual statistics for the selected GMM, plus an estimate of the hypervolume associated with the noise component. The classification obtained is shown in Figure 7.2b, which reflcets the generating process for the data quite well. The cross-tabulation of the estimated classification and the known data labels shows that most of the clutter is correctly identified, as is the true minefield. The sensitivity is $= 743/754 = 98.54\%$, and the specificity is $= (156 + 144)/350 = 85.71\%$.

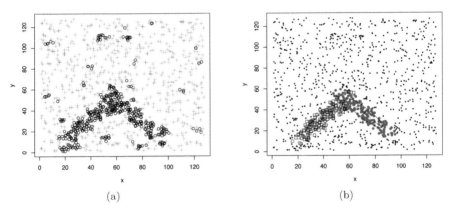

(a)   (b)

FIGURE 7.2: Model-based clustering of a simulated minefield with noise: (a) initial denoised minefield data; (b) model-based clustering partition obtained with the noise component.

By default **mclust** uses the function hypvol() to compute the hypervolume V in equation (7.1). This gives a simple approximation to the hypervolume of a multivariate dataset by taking the minimum of the volume hyperrectangle (box) containing the observed data and the box obtained from principal components. In other words, the estimate is the minimum of the Cartesian product of the variable intervals and the Cartesian product of the corresponding projection onto principal components.

```
hypvol(X)
## [1] 16022
```

Another option is to compute the *ellipsoid hull*, namely the ellipsoid of minimal volume such that all observed points lie either inside or on the boundary of the ellipsoid. This method is available via the function `ellipsoidhull()` from the **cluster** R package (Maechler et al., 2022):

```
library("cluster")
ehull <- ellipsoidhull(as.matrix(X))
volume(ehull)
## [1] 23150
modNoise.ehull <- Mclust(X, Vinv = 1/volume(ehull),
                         initialization = list(noise = (nnc$z == 0)))
summary(modNoise.ehull)
## ------------------------------------------------------
## Gaussian finite mixture model fitted by EM algorithm
## ------------------------------------------------------
##
## Mclust VVE (ellipsoidal, equal orientation) model with 3 components and
## a noise term:
##
## log-likelihood    n df    BIC    ICL
##          -10777 1104 17 -21673 -22618
##
## Clustering table:
##   1   2   3   0
## 187 192 302 423
tab <- table(chevron$class, modNoise.ehull$classification)
addmargins(tab, 2)
##
##             0   1   2   3 Sum
##    data     0 164 164  22 350
##    noise  423  23  28 280 754
```

In this case the number of clusters identified is too large, yielding a poor recovery of the noise component (sensitivity $= 423/754 = 56.1\%$, and specificity $= (164 + 164 + 22)/350 = 100\%$). This behavior is a consequence of the ellipsoid hull over-estimating the data volume (to be expected because the noise is distributed over a square, so that a larger volume ellipsoid is needed to contain all the data points). The hyperrectangle approach discussed above would in all cases be a better approximation to the data volume than the ellipsoid hull. However, the volume of the convex hull of the data better approximates the data volume than either of these approaches.

The following code obtains the volume of the convex hull for the bivariate minefield simulated data, using the `convhulln()` function from package **geometry** (Habel et al., 2022):

```
library("geometry")
chull <- convhulln(X, options = "FA")
chull$vol
## [1] 15828
```

The reciprocal of this value may be used as input for the argument Vinv of function Mclust():

```
modNoise.chull <- Mclust(X, Vinv = 1/chull$vol,
                         initialization = list(noise = (nnc$z == 0)))
summary(modNoise.chull)
## ----------------------------------------------------
## Gaussian finite mixture model fitted by EM algorithm
## ----------------------------------------------------
##
## Mclust EEV (ellipsoidal, equal volume and shape) model with 2
## components and a noise term:
##
## log-likelihood     n df    BIC    ICL
##           -10515 1104 11 -21108 -21487
##
## Clustering table:
##    1   2   0
## 157 149 798
tab <- table(chevron$class, modNoise.chull$classification)
addmargins(tab, 2)
##
##            0   1   2 Sum
##    data   55 153 142 350
##    noise 743   4   7 754
```

For this dataset, the default value of the hypervolume and that computed using convhulln() are quite close, so the corresponding estimated models provide essentially the same fit.

An issue with the convex hull is that it may not be practical to compute in high dimensions. Moreover, implementations without the option to return the logarithm could result in overflow or underflow depending on the scaling of the data.

Other strategies for obtaining an initial noise estimate could be adopted. The function cov.nnve() in the contributed R package **covRobust** (Wang et al., 2017) is an implementation of robust covariance estimation via nearest neighbor cleaning (Wang and Raftery, 2002). An example of its usage is the following:

```
library("covRobust")
nnve <- cov.nnve(X, k = 5)
table(nnve$classification)
##
##    0    1
##    5 1099
```

For the case of $k = 5$ nearest neighbors, only $5/1104 \approx 0.05\%$ of data is classified as noise. A relatively good clustering model is nevertheless obtained with this initial estimate of the noise:

```
modNoise.nnve <- Mclust(X, initialization =
                        list(noise = (nnve$classification == 0)))
summary(modNoise.nnve$BIC)
## Best BIC values:
##             EEV,2         EVE,2        EVV,2
## BIC      -21128 -21128.20252 -21134.8273
## BIC diff      0      -0.36423     -6.9891
summary(modNoise.nnve)
## ----------------------------------------------------
## Gaussian finite mixture model fitted by EM algorithm
## ----------------------------------------------------
##
## Mclust EEV (ellipsoidal, equal volume and shape) model with 2
## components and a noise term:
##
##  log-likelihood    n df    BIC    ICL
##          -10525 1104 11 -21128 -21509
##
## Clustering table:
##   1    2    0
## 160 150 794
addmargins(table(chevron$class, modNoise.nnve$classification), 2)
##
##            0    1    2 Sum
##   data    51 153 146 350
##   noise 743    7    4 754
```

Although the final model is somewhat different from the one previously selected, this one also has 2 components, and the classification of data as signal or noise is almost identical (sensitivity $= 743/754 = 98.54\%$, and specificity $= (153+146)/350 = 85.43\%$). By increasing the number of nearest neighbors k for the initial noise estimate, the results for the cov.nnve() denoising approach those obtained with NNclean.

EXAMPLE 7.2: Clustering with outliers on simulated data

Peel and McLachlan (2000) developed a simulation-based example in the context of fitting of mixtures of (multivariate) t distributions to model data containing observations with longer than normal tails or atypical observations. The data used in their example is generated as follows:

```
(Sigma <- array(c(2, 0.5, 0.5, 0.5, 1, 0, 0, 0.1, 2, -0.5, -0.5, 0.5),
                dim = c(2, 2, 3)))
## , , 1
##
##      [,1] [,2]
## [1,]  2.0  0.5
## [2,]  0.5  0.5
##
## , , 2
##
##      [,1] [,2]
## [1,]    1  0.0
## [2,]    0  0.1
##
## , , 3
##
##      [,1] [,2]
## [1,]  2.0 -0.5
## [2,] -0.5  0.5
var.decomp <- sigma2decomp(Sigma)
str(var.decomp)
## List of 7
## $ sigma : num [1:2, 1:2, 1:3] 2 0.5 0.5 0.5 1 0 0 0.1 2 -0.5 ...
## $ d : int 2
## $ modelName : chr "VVI"
## $ G : int 3
## $ scale : num [1:3] 0.866 0.316 0.866
## $ shape : num [1:2, 1:3] 2.484 0.403 3.162 0.316 2.484 ...
## $ orientation: num [1:2, 1:2] -0.957 -0.29 0.29 -0.957
par <- list(pro = c(1/3, 1/3, 1/3),
            mean = cbind(c(0, 3), c(3, 0), c(-3, 0)),
            variance = var.decomp)
data <- sim(par$variance$modelName, parameters = par, n = 200)
noise <- matrix(runif(100, -10, 10), nrow = 50, ncol = 2)
X <- rbind(data[, 2:3], noise)
cluster <- c(data[, 1], rep(0, 50))
clPairs(X, ifelse(cluster == 0, 0, 1),
        colors = "black", symbols = c(16, 1), cex = c(0.5, 1))
```

In the code above we used the function sim() to simulate 200 observations from a GMM with the specified mixing probabilities pro, component means mean, and variance decomposition var.decomp obtained using the function sigma2decomp() (see Section 7.4). Then, we added 50 outlying observations from a uniform distribution on $(-10, 10)$. The simulated data is shown in Figure 7.3a.

We apply the same procedure adopted for noisy data in Example 7.1. We get an initial estimate of the outlying observations with NNclean(), and then we apply Mclust() to all the data using this estimate of noise in the initialization step:

```
nnc <- NNclean(X, k = 5)
modNoise <- Mclust(X, initialization = list(noise = (nnc$z == 0)))
summary(modNoise$BIC)
## Best BIC values:
##             VEI,3       VEE,3       VVI,3
## BIC      -2243.8 -2249.2536 -2253.0499
## BIC diff     0.0     -5.4888     -9.2851
summary(modNoise, parameters = TRUE)
## ----------------------------------------------------
## Gaussian finite mixture model fitted by EM algorithm
## ----------------------------------------------------
##
## Mclust VEI (diagonal, equal shape) model with 3 components and a noise
## term:
##
## log-likelihood    n df     BIC     ICL
##         -1083.2 250 14 -2243.8 -2267.4
##
## Clustering table:
## 1  2  3  0
## 66 63 74 47
##
## Mixing probabilities:
##        1       2       3       0
## 0.25532 0.24244 0.28647 0.21578
##
## Means:
##        [,1]     [,2]     [,3]
## x1 2.968281 -0.40142 -3.13151
## x2 0.022729  3.01644  0.13098
##
## Variances:
## [,,1]
##        x1       x2
## x1 0.7887 0.00000
```

```
## x2 0.0000 0.12971
## [,,2]
##        x1       x2
## x1 1.896 0.00000
## x2 0.000 0.31181
## [,,3]
##         x1       x2
## x1 1.9577 0.00000
## x2 0.0000 0.32195
##
## Hypervolume of noise component:
## 377.35
```

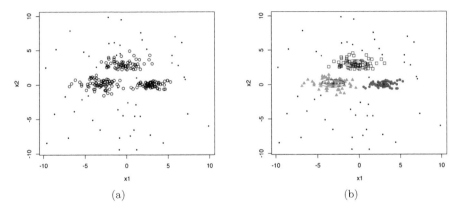

(a)   (b)

FIGURE 7.3: Model-based clustering of a bivariate mixture of normal distributions with outlying observations: (a) the simulated data; (b) model-based clustering partition obtained with the noise component. The small black points represent the simulated noise in (a) and the estimated noise in (b).

The clusters identified are shown in Figure 7.3b, and the following code can be used to plot the result and obtain the confusion matrix:

```
plot(modNoise, what = "classification")
table(cluster, Classification = modNoise$classification)
##         Classification
## cluster  0  1  2  3
##       0 45  0  3  2
##       1  2  0 60  0
##       2  0 64  0  0
##       3  0  2  0 72
```

This shows that the three groups (labeled "1", "2", "3") and the outlying observations (labeled "0") have been well identified.

7.2   Using a Prior for Regularization

Maximum likelihood estimation for Gaussian mixtures using the EM algorithm may fail as the result of singularities or degeneracies. These problems can arise depending on the data and the starting values for the EM algorithm, as well as the model parameterizations and number of components specified for the mixture.

To address these situations, Fraley and Raftery (2007) proposed replacing the maximum likelihood estimate (MLE) by a maximum a posteriori (MAP) estimate, also estimated by the EM algorithm. A highly dispersed proper conjugate prior is used, containing a small fraction of one observation's worth of information. The resulting method avoids degeneracies and singularities, but when these are not present it gives similar results to the standard maximum likelihood method. The prior also has the effect of smoothing noisy behavior of the BIC, which is often observed in conjunction with instability in estimation.

For univariate data, a Gaussian prior on the mean (conditional on the variance) is used:

$$\mu \mid \sigma^2 \sim \mathcal{N}(\mu_{\mathcal{P}}, \sigma^2/\kappa_{\mathcal{P}})$$

$$\propto \left(\sigma^2\right)^{-\frac{1}{2}} \exp\left\{-\frac{\kappa_{\mathcal{P}}}{2\sigma^2}\left(\mu - \mu_{\mathcal{P}}\right)^2\right\}, \tag{7.2}$$

and an inverse gamma prior on the variance:

$$\sigma^2 \sim \text{inverseGamma}(\nu_{\mathcal{P}}/2, \varsigma_{\mathcal{P}}^2/2)$$

$$\propto \left(\sigma^2\right)^{-\frac{\nu_{\mathcal{P}}+2}{2}} \exp\left\{-\frac{\varsigma_{\mathcal{P}}^2}{2\sigma^2}\right\}. \tag{7.3}$$

For multivariate data, a Gaussian prior on the mean (conditional on the covariance matrix) is used:

$$\boldsymbol{\mu} \mid \boldsymbol{\Sigma} \sim \mathcal{N}(\boldsymbol{\mu}_{\mathcal{P}}, \boldsymbol{\Sigma}/\kappa_{\mathcal{P}})$$

$$\propto |\boldsymbol{\Sigma}|^{-\frac{1}{2}} \exp\left\{-\frac{\kappa_{\mathcal{P}}}{2}\operatorname{tr}\left((\boldsymbol{\mu} - \boldsymbol{\mu}_{\mathcal{P}})^{\top}\boldsymbol{\Sigma}^{-1}(\boldsymbol{\mu} - \boldsymbol{\mu}_{\mathcal{P}})\right)\right\}, \tag{7.4}$$

together with an inverse Wishart prior on the covariance matrix:

$$\boldsymbol{\Sigma} \sim \text{inverseWishart}(\nu_{\mathcal{P}}, \boldsymbol{\Lambda}_{\mathcal{P}})$$

$$\propto |\boldsymbol{\Sigma}|^{-\frac{\nu_{\mathcal{P}}+d+1}{2}} \exp\left\{-\frac{1}{2}\operatorname{tr}\left(\boldsymbol{\Sigma}^{-1}\boldsymbol{\Lambda}_{\mathcal{P}}^{-1}\right)\right\}. \tag{7.5}$$

The hyperparameters $\mu_{\mathcal{P}}$ (in the univariate case) or $\boldsymbol{\mu}_{\mathcal{P}}$ (in the multivariate case), $\kappa_{\mathcal{P}}$, and $\nu_{\mathcal{P}}$ are called the *mean, shrinkage*, and *degrees of freedom*, respectively. Parameters $\varsigma_{\mathcal{P}}^2$ (a scalar, in the univariate case) and $\boldsymbol{\Lambda}_{\mathcal{P}}$ (a matrix, in the multivariate case) are the *scale* of the prior distribution. These priors are called *conjugate priors* for the normal distribution because the posterior can be expressed as the product of a normal distribution and an inverse gamma or Wishart distribution.

When a prior is included, a modified version of the BIC must be used to select the number of mixture components and the model parameterization. The BIC is computed as described in Section 2.3.1, but with the log-likelihood evaluated at the MAP instead of at the MLE. Further details on model-based clustering using prior distributions can be found in Fraley and Raftery (2007).

7.2.1 Adding a Prior in mclust

Although maximum likelihood estimation is the default, functions such as Mclust() and mclustBIC() have a prior argument that allows specification of a conjugate prior on the means and variances of the type described above. **mclust** provides two functions to facilitate inclusion of a prior:

- priorControl() supplies input for the prior argument. It gives the name of the function that defines the desired conjugate prior and specifies values for its arguments.

- defaultPrior() is the default function named in priorControl(), as well as a template for specifying conjugate priors in **mclust**.

The defaultPrior() function has the following arguments:

data A numeric vector, matrix, or data frame of observations.

G An integer value specifying the number of mixture components.

modelName A character string indicating the model to be fitted. Note that in the multivariate case only 10 out of 14 models are available, as indicated in Table 3.1.

The following optional arguments can also be specified:

mean A single value $\mu_{\mathcal{P}}$ or a vector $\boldsymbol{\mu}_{\mathcal{P}}$ of values for the mean parameter of the prior. By default, the sample mean of each variable is used.

shrinkage The shrinkage parameter $\kappa_{\mathcal{P}}$ for the prior on the mean. By default, shrinkage = 0.01. The posterior mean

$$\frac{n_k \bar{\boldsymbol{x}}_k + \kappa_{\mathcal{P}} \boldsymbol{\mu}_{\mathcal{P}}}{\kappa_{\mathcal{P}} + n_k},$$

can be viewed as adding $\kappa_{\mathcal{P}}$ observations with value $\boldsymbol{\mu}_{\mathcal{P}}$ to each group in the data. The default value was determined by experimentation; values close to and larger than 1 caused large perturbations in the modeling in cases where there were no missing BIC values without the prior. The value $\kappa_{\mathcal{P}} = 0.01$ resulted in BIC curves that appeared to be smooth extensions of their counterparts without the prior. By setting `shrinkage = 0` or `shrinkage = NA` no prior is assumed for the mean.

dof The degrees of freedom $\nu_{\mathcal{P}}$ for the prior on the variance. By analogy to the univariate case, the marginal prior distribution of $\boldsymbol{\mu}$ is a t distribution centered at $\boldsymbol{\mu}_{\mathcal{P}}$ with $\nu_{\mathcal{P}} - d + 1$ degrees of freedom. The mean of this distribution is $\boldsymbol{\mu}_{\mathcal{P}}$ provided that $\nu_{\mathcal{P}} > d$, and it has a finite covariance matrix provided $\nu_{\mathcal{P}} > d + 1$ (see, for example Schafer, 1997). By default, $\nu_{\mathcal{P}} = d + 2$, the smallest integer value for the degrees of freedom that gives a finite covariance matrix.

scale The scale parameter for the prior on the variance.
For univariate models and multivariate spherical or diagonal models, it is computed by default as

$$\varsigma_{\mathcal{P}}^2 = \frac{\text{tr}(\boldsymbol{S})/d}{G^{2/d}},$$

which is the average of the diagonal elements of the empirical covariance matrix $\boldsymbol{S}$ of the data, divided by the square of the number of components to the $1/d$ power. This is roughly equivalent to partitioning the range of the data into G intervals of fairly equal size.
For multivariate ellipsoidal models, it is computed by default as

$$\boldsymbol{\Lambda}_{\mathcal{P}} = \frac{\boldsymbol{S}}{G^{2/d}},$$

the covariance matrix, divided by the square of the number of components to the $1/d$ power.

EXAMPLE 7.3: Density estimation with a prior on the Galaxies data

Consider the data on the estimated velocity (km/sec) of 82 galaxies in the Corona Borealis region from Roeder (1990), available in the **MASS** package (Venables and Ripley, 2013; Ripley, 2022). The interest lies in the identification of multimodality as evidence for clustering structures in the pattern of expansion.

```
data("galaxies", package = "MASS")
# now fix a typographical error in the data
# see help("galaxies", package = "MASS")
galaxies[78] <- 26960
galaxies <- galaxies / 1000
```

We start the analysis with no prior, but using multiple random starts and retaining the best estimated fit in terms of BIC:

```
BIC <- NULL
for(i in 1:50)
{
  BIC0 <- mclustBIC(galaxies, verbose = FALSE,
               initialization = list(hcPairs = hcRandomPairs(galaxies)))
  BIC  <- mclustBICupdate(BIC, BIC0)
}
summary(BIC, k = 5)
## Best BIC values:
##               V,6         V,3         V,5         V,4         E,6
## BIC       -440.06 -442.2177 -442.5948 -443.8970 -447.4629
## BIC diff    0.00    -2.1543    -2.5314    -3.8335    -7.3994
plot(BIC)

mod <- densityMclust(galaxies, x = BIC)
summary(mod, parameters = TRUE)
## ----------------------------------------------------------
## Density estimation via Gaussian finite mixture modeling
## ----------------------------------------------------------
##
## Mclust V (univariate, unequal variance) model with 6 components:
##
##  log-likelihood  n df      BIC    ICL
##          -182.57 82 17 -440.06 -447.6
##
## Mixing probabilities:
##        1        2        3        4        5        6
## 0.403204 0.085366 0.024390 0.024355 0.036585 0.426099
##
## Means:
##       1       2       3       4       5       6
## 19.7881  9.7101 16.1270 26.9775 33.0443 22.9162
##
## Variances:
##          1          2          3          4          5          6
## 0.45348851 0.17851527 0.00184900 0.00030625 0.84956356 1.45114183
plot(mod, what = "density", data = galaxies, breaks = 11)
rug(galaxies)
```

The model with the largest BIC has 6 components, each with a different variance. Components with very small mixing weights correspond to narrow spikes in the density estimate (see Figure 7.4), suggesting them to be spurious.

The previous analysis can be modified by including the prior as follows:

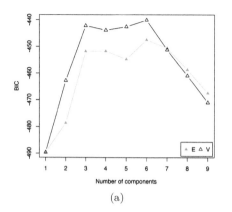
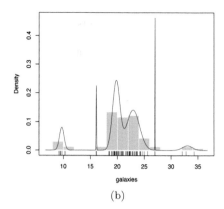

(a)   (b)

FIGURE 7.4: BIC plot without the prior (a) and density estimate for the (V,6) model selected according to BIC for the galaxies dataset.

```
BICp <- NULL
for(i in 1:50)
{
  BIC0p <- mclustBIC(galaxies, verbose = FALSE,
                     prior = priorControl(),
           initialization = list(hcPairs = hcRandomPairs(galaxies)))
  BICp  <- mclustBICupdate(BICp, BIC0p)
}
summary(BICp, k = 5)
## Best BIC values:
##                V,3       V,4        E,6        E,7        E,3
## BIC      -443.96 -445.153 -447.6407 -450.6451 -452.0662
## BIC diff    0.00   -1.196   -3.6833   -6.6878   -8.1088
plot(BICp)

modp <- densityMclust(galaxies, x = BICp)
summary(modp, parameters = TRUE)
## -----------------------------------------------------------
## Density estimation via Gaussian finite mixture modeling
## -----------------------------------------------------------
##
## Mclust V (univariate, unequal variance) model with 3 components:
##
## Prior: defaultPrior()
##
##  log-likelihood  n df      BIC      ICL
##         -204.35 82  8 -443.96 -443.96
```

```
##
## Mixing probabilities:
##        1        2        3
## 0.085366 0.036585 0.878050
##
## Means:
##       1       2       3
##   9.726  33.004  21.404
##
## Variances:
##        1        2        3
## 0.36949 0.70599 4.51272
plot(modp, what = "density", data = galaxies, breaks = 11)
rug(galaxies)
```

Figure 7.5 shows the BIC trace and the density estimate for the selected model. By including the prior, BIC fairly clearly chooses the model with three components and unequal variances. This is in agreement with other analyses reported in the literature (Roeder and Wasserman, 1997; Fraley and Raftery, 2007).

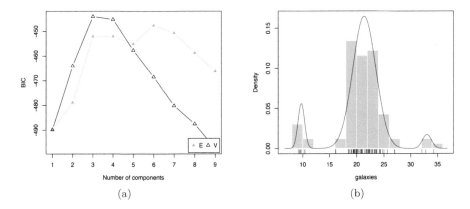

(a)   (b)

FIGURE 7.5: BIC plot with the prior (a) and density estimate for the (V,3) model selected according to the BIC (based on the MAP, not the MLE) for the galaxies dataset.

As explained in Section 7.2.1, the function priorControl() is provided for specifying a prior and its parameters. When used with its default settings, it specifies another function called defaultPrior() which can serve as a template for alternative priors. An example of the result of a call to defaultPrior() for the final model selected is shown below:

```
defaultPrior(galaxies, G = 3, modelName = "V")
## $shrinkage
## [1] 0.01
##
## $mean
## [1] 20.831
##
## $dof
## [1] 3
##
## $scale
## [1] 2.3187
```

EXAMPLE 7.4: Clustering with a prior on the Italian olive oils data

Consider the olive oils dataset presented in Example 4.8.

```
data("olive", package = "pgmm")
# recode of labels for Region and Area
Regions <- c("South", "Sardinia", "North")
Areas <- c("North Apulia", "Calabria", "South Apulia", "Sicily",
           "Inland Sardinia", "Coastal Sardinia", "East Liguria",
           "West Liguria", "Umbria")
olive$Region <- factor(olive$Region, levels = 1:3, labels = Regions)
olive$Area <- factor(olive$Area, levels = 1:9, labels = Areas)
with(olive, table(Area, Region))
##                   Region
## Area             South Sardinia North
##    North Apulia     25        0     0
##    Calabria         56        0     0
##    South Apulia    206        0     0
##    Sicily           36        0     0
##    Inland Sardinia   0       65     0
##    Coastal Sardinia  0       33     0
##    East Liguria      0        0    50
##    West Liguria      0        0    50
##    Umbria            0        0    51
```

We model the data on the standardized scale as follows:

```
X <- scale(olive[, 3:10])

BIC <- mclustBIC(X, G = 1:15)
summary(BIC)
```

```
## Best BIC values:
##               VVE,14      VVE,11      VVE,13
## BIC       -4461.6 -4471.3336 -4480.081
## BIC diff      0.0    -9.6941   -18.442
plot(BIC, legendArgs = list(x = "bottomright", ncol = 5))

BICp <- mclustBIC(X, G = 1:15, prior = priorControl())
summary(BICp)
## Best BIC values:
##              VVV,6     VVV,7     VVV,5
## BIC      -5446.7 -5584.27 -5590.59
## BIC diff     0.0  -137.55  -143.86
plot(BICp, legendArgs = list(x = "bottomright", ncol = 5))
```

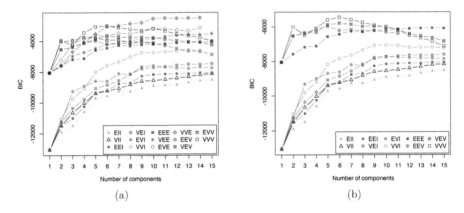

FIGURE 7.6: BIC plot without prior (a) and with prior (b) for the olive oil
data.

The BIC plots without and with a prior are shown in Figure 7.6. Without a
prior, the model selected has a large number of mixture components, and a
fairly poor clustering accuracy:

```
mod <- Mclust(X, x = BIC)
summary(mod)
## ----------------------------------------------------
## Gaussian finite mixture model fitted by EM algorithm
## ----------------------------------------------------
##
## Mclust VVE (ellipsoidal, equal orientation) model with 14 components:
##
```

```
## log-likelihood   n  df    BIC    ICL
##        -1389.6 572 265 -4461.6 -4504.4
##
## Clustering table:
##  1  2  3  4  5  6  7  8  9 10 11 12 13 14
## 29 51 34 65 54 90 66 32 11 29 16 41 38 16
table(Region = olive$Region, cluster = mod$classification)
##            cluster
## Region      1  2  3  4  5  6  7  8  9 10 11 12 13 14
##   South    29 51 34 65 54 90  0  0  0  0  0  0  0  0
##   Sardinia  0  0  0  0  0  0 66 32  0  0  0  0  0  0
##   North     0  0  0  0  0  0  0  0 11 29 16 41 38 16
adjustedRandIndex(olive$Region, mod$classification)
## [1] 0.24185
table(Area = olive$Area, cluster = mod$classification)
##                   cluster
## Area               1  2  3  4  5  6  7  8  9 10 11 12 13 14
##   North Apulia    23  0  1  1  0  0  0  0  0  0  0  0  0  0
##   Calabria         0 48  2  6  0  0  0  0  0  0  0  0  0  0
##   South Apulia     0  0  6 57 53 90  0  0  0  0  0  0  0  0
##   Sicily           6  3 25  1  1  0  0  0  0  0  0  0  0  0
##   Inland Sardinia  0  0  0  0  0  0 65  0  0  0  0  0  0  0
##   Coastal Sardinia 0  0  0  0  0  0  1 32  0  0  0  0  0  0
##   East Liguria     0  0  0  0  0  0  0  0  0  1  4 41  4  0
##   West Liguria     0  0  0  0  0  0  0  0  0  0  0  0 34 16
##   Umbria           0  0  0  0  0  0  0  0 11 28 12  0  0  0
adjustedRandIndex(olive$Area, mod$classification)
## [1] 0.54197
```

The model selected when the prior is included has 6 components with unconstrained variances and a much improved clustering accuracy:

```
modp <- Mclust(X, x = BICp)
summary(modp)
## ------------------------------------------------------
## Gaussian finite mixture model fitted by EM algorithm
## ------------------------------------------------------
##
## Mclust VVV (ellipsoidal, varying volume, shape, and orientation) model
## with 6 components:
##
## Prior: defaultPrior()
##
## log-likelihood   n  df    BIC    ICL
##        -1869.4 572 269 -5446.7 -5451.4
```

```
##
## Clustering table:
##   1   2   3   4   5   6
## 127 196  98  36  58  57
table(Region = olive$Region, cluster = modp$classification)
##            cluster
## Region        1   2   3   4   5   6
##    South     127 196   0   0   0   0
##    Sardinia    0   0  98   0   0   0
##    North       0   0   0  36  58  57
adjustedRandIndex(olive$Region, modp$classification)
## [1] 0.56313
table(Area = olive$Area, cluster = modp$classification)
##                    cluster
## Area                  1   2   3   4   5   6
##    North Apulia      25   0   0   0   0   0
##    Calabria          56   0   0   0   0   0
##    South Apulia      10 196   0   0   0   0
##    Sicily            36   0   0   0   0   0
##    Inland Sardinia    0   0  65   0   0   0
##    Coastal Sardinia   0   0  33   0   0   0
##    East Liguria       0   0   0   0  43   7
##    West Liguria       0   0   0   0   0  50
##    Umbria             0   0   0  36  15   0
adjustedRandIndex(olive$Area, modp$classification)
## [1] 0.78261
```

The clusters identified correspond approximately to the areas of origin of the olive oil samples. One cluster encompasses both the inland and coastal areas of Sardinia, and two clusters include the southern regions, with one almost completely composed of the south of Apulia. Another three clusters correspond to northern regions: one for all the olive oils from west Liguria and a few from east Liguria, one mostly composed of east Liguria and a few from Umbria, and the last cluster composed only of olive oils from Umbria.

7.3 Non-Gaussian Clusters from GMMs

Cluster analysis aims to identify groups of similar observations that are relatively separated from each other. In the model-based approach to clustering, data is modeled by a finite mixture of density functions belonging to a given parametric class. Each component is typically associated with a group or cluster. This framework can be extended to non-Gaussian clusters by modeling them with more than one mixture component. This problem was addressed by Baudry et al. (2010), who proposed a method based on an entropy criterion for hierarchically combining mixture components to form clusters. Hennig (2010) also discussed hierarchical merging methods based on unimodality and misclassification probabilities. Another approach based on the identification of cluster cores corresponding to regions of high density is introduced by Scrucca (2016).

7.3.1 Combining Gaussian Mixture Components for Clustering

The function `clustCombi()` provides a means for obtaining models with clusters represented by multiple mixture components, as illustrated below.

EXAMPLE 7.5: Merging Gaussian mixture components on simulated data

Consider the following simulated two-dimensional dataset with overlapping components as discussed in Baudry et al. (2010, Sec. 4.1).

```
data("Baudry_etal_2010_JCGS_examples", package = "mclust")
plot(ex4.1)
```

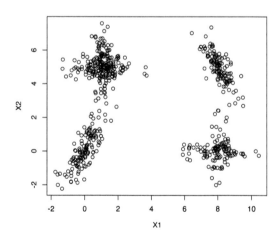

FIGURE 7.7: The ex4.1 simulated dataset.

A sample of 600 observations was generated from a mixture of six Gaussian components, although it would be reasonable to conclude from the spatial distribution of the data that there are only four clusters. The data is shown in Figure 7.7. Two of the clusters are formed from overlapping axis-aligned components (diagonal covariance matrices), while the remaining two clusters are elliptical but not axis-aligned.

The estimated model maximizing the BIC criterion is the following:

```
mod_ex4.1 <- Mclust(ex4.1)
summary(mod_ex4.1)
## ----------------------------------------------------
## Gaussian finite mixture model fitted by EM algorithm
## ----------------------------------------------------
##
## Mclust EEV (ellipsoidal, equal volume and shape) model with 6
## components:
##
##  log-likelihood   n df      BIC     ICL
##         -1957.8 600 25 -4075.5 -4248.4
##
## Clustering table:
##    1   2   3   4   5   6
##   78 122 121 107 132  40
```

The function clustCombi(), following the methodology described in Baudry et al. (2010), combines the mixture components hierarchically according to an entropy criterion. The estimated models with numbers of classes ranging from a single class to the number of components selected by BIC are returned as follows:

```
CLUSTCOMBI <- clustCombi(mod_ex4.1)
summary(CLUSTCOMBI)
## ----------------------------------------------------
## Combining Gaussian mixture components for clustering
## ----------------------------------------------------
##
## Mclust model name: EEV
## Number of components: 6
##
## Combining steps:
##
##   Step | Classes combined at this step | Class labels after this step
## -------|-------------------------------|-----------------------------
##    0   |             ---               | 1 2 3 4 5 6
##    1   |            3 & 4              | 1 2 3 5 6
##    2   |            1 & 6              | 1 2 3 5
##    3   |            3 & 5              | 1 2 3
##    4   |            1 & 2              | 1 3
##    5   |            1 & 3              | 1
```

The summary output shows the mixture model EEV with 6 components selected by the BIC criterion, followed by information describing the combining steps. The hierarchy of combined components can be displayed graphically as follows (see Figure 7.8):

```
par(mfrow = c(3, 2), mar = c(4, 4, 3, 1))
plot(CLUSTCOMBI, ex4.1, what = "classification")
```

The process of merging mixture components can also be represented using tree diagrams:

```
plot(CLUSTCOMBI, what = "tree")
plot(CLUSTCOMBI, what = "tree", type = "rectangle", yaxis = "step")
```

Figure 7.9 shows the resulting tree obtained using the default arguments, type = "triangle" and yaxis = "entropy". The first argument controls the type of dendrogram to plot; two options are available, namely, "triangle" and "rectangle". The second argument controls the quantity to appear on the y-axis; by default one minus the normalized entropy at each merging step is used. The option yaxis = "step" is also available to simply represent heights as successive steps, as shown in Figure 7.9b.

The clustering structures obtained with clustCombi() can be compared on substantive grounds or by selecting the number of clusters via a piecewise linear regression fit to the (possibly rescaled) "entropy plot". An automatic procedure to help the user to select the optimal number of clusters using the above methodology based on the entropy is available through the function clustCombiOptim(). For example:

```
optimClust <- clustCombiOptim(CLUSTCOMBI, reg = 2, plot = TRUE)
str(optimClust)
## List of 3
## $ numClusters.combi: int 4
## $ z.combi : num [1:600, 1:4] 1.00 1.00 5.78e-04 1.85e-42 3.85e-35 ...
## $ cluster.combi : num [1:600] 1 1 2 3 3 4 4 3 3 2 ...
```

produces the entropy plot in Figure 7.10 and returns a list containing the number of clusters (numClusters.combi), the probabilities (z.combi) obtained by summing posterior conditional probabilities over merged components, and the clustering labels (cluster.combi) obtained by merging the mixture components. The optional argument reg is used to specify the number of segments to use in the piecewise linear regression fit. Possible choices are 2 (the default, for a model with two segments or one change point) and 3 (for a model with three segments or two change points). The entropy plot clearly suggests a four-cluster solution.

Finally, we note that an alternative method for merging Gaussian mixtures estimated using **mclust** has been proposed by Hennig (2010) and it is available through the function mergenormals() contained in the **fpc** R package (Hennig, 2020).

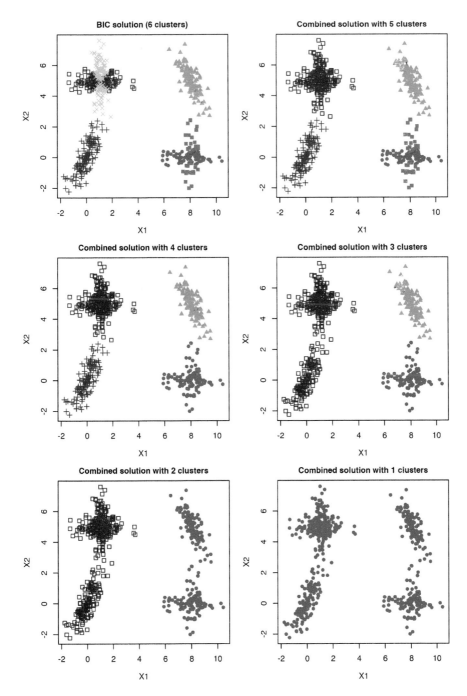

FIGURE 7.8: Hierarchy of mixture component combinations for the ex4.1 dataset.

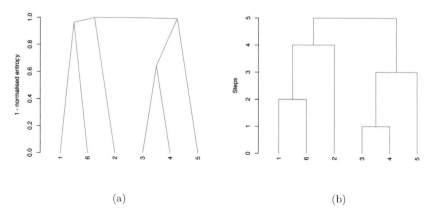

(a)   (b)

FIGURE 7.9: Dendrogram tree plots of the merging structure of mixture component combinations for the simulated ex4.1 dataset.

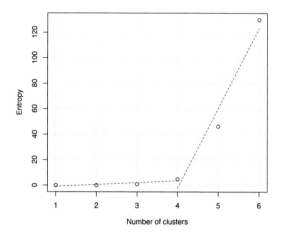

FIGURE 7.10: Entropy plot for selecting the final clustering by merging mixture components for the simulated ex4.1 dataset.

7.3.2 Identifying Connected Components in GMMs

A limitation of approaches that successively combine mixture components is that, once merged, components cannot subsequently be reallocated to different clusters at a later stage of the algorithm.

Among the several definitions of "cluster" available in the literature, Hartigan (1975, p. 205) proposed that "[...] clusters may be thought of as regions of high density separated from other such regions by regions of low density." From this perspective, Scrucca (2016) defined high density clusters as connected components of density level sets. Starting with a density estimate obtained from a Gaussian finite mixture model, cluster cores — those data points which form the core of the clusters — are obtained from the connected components at a given density level. A mode function gives the number of connected components as the level is varied. Once cluster cores are identified, the remaining observations are allocated to those cluster cores for which the probability of cluster membership is the highest. Details are provided in Scrucca (2016) and implemented in the gmmhd() function available in **mclust**. A non-parametric variant proposed by Azzalini and Menardi (2014) is implemented in **pdfCluster** (Giovanna and Adelchi, 2022).

EXAMPLE 7.6: Identifying high density clusters on the Old Faithful data

As a first example, consider the Old Faithful data described in Example 3.4.

```
data("faithful", package = "datasets")
mod <- Mclust(faithful)
clPairs(faithful, mod$classification)
plot(as.densityMclust(mod), what = "density", add = TRUE)
```

Figure 7.11a shows the clustering partition obtained from the best estimated GMM according to BIC. The model selected is (EEE,3), a mixture of three components with common full covariance matrix. The portion of the data with high values for both waiting and eruptions is fitted with two Gaussian components. However, there appear to be two separate groups of points in the data as a whole, and the corresponding bivariate density estimate shown in Figure 7.11a indicates the presence of two separate regions of high density.

The GMMHD approach (Scrucca, 2016) briefly outlined at the beginning of this section is implemented in the gmmhd() function available in **mclust**. This function requires as input an object returned by Mclust() or densityMclust() as the initial GMM. Many other input parameters can be set, and the interested reader should consult the documentation available through help("gmmhd").

We apply the GMMHD algorithm to our example by invoking function gmmhd() and producing its summary output as follows:

```
GMMHD <- gmmhd(mod)
summary(GMMHD)
## ------------------------------------------------------------
## GMM with high-density connected components for clustering
## ------------------------------------------------------------
##
## Initial model:  Mclust (EEE,3)
##
## Cluster cores:
##    1    2 <NA>
##  166   91   15
##
## Final clustering:
##    1    2
##  178   94
```

Starting with the density estimate from the initial (EEE,3) Gaussian mixture model, gmmhd() computes the empirical mode function and the connected components at different density levels, which are then used to identify cluster cores. Figure 7.11b plots the number of modes as a function of the proportion of data points above a threshold density level. There is a clear indication of a bimodal distribution. Figure 7.11c shows the points assigned to the cluster cores, as well as the remaining data points not initially allocated.

The summary above shows that 166 and 91 observations have been assigned to the two cluster cores, and that 15 points were initially unlabeled. Figure 7.11d shows the final clustering obtained after the unlabeled data have also been allocated to the cluster cores. Figures 7.11b–7.11d can be produced with the following commands:

```
plot(GMMHD, what = "mode")
plot(GMMHD, what = "cores")
plot(GMMHD, what = "clusters")
```

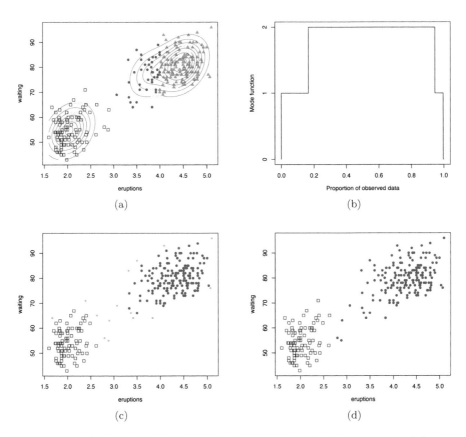

FIGURE 7.11: Identification of connected components for the Old Faithful data:
(a) clustering and density contours for the initial GMM; (b) mode function;
(c) cluster cores (marked as • and □) and unlabeled data (marked as ○); (d)
final clustering obtained after unlabeled data have been allocated.

7.4   Simulation from Mixture Densities

Given the parameters of a mixture model, data can be simulated from that model for evaluation and verification. The **mclust** function sim() enables simulation from all of the structured GMMs supported. Besides the model, sim() allows a seed to be input for reproducibility.

EXAMPLE 7.7: Simulating data from GMM estimated from the Old Faithful data

In the example below we simulate two different datasets of the same size as the faithful dataset from the model returned by Mclust():

```
mod <- Mclust(faithful)
sim0 <- sim(modelName = mod$modelName,
            parameters = mod$parameters,
            n = nrow(faithful), seed = 0)
sim1 <- sim(modelName = mod$modelName,
            parameters = mod$parameters,
            n = nrow(faithful), seed = 1)
```

The results can be plotted as follows (graphs not shown):

```
xlim <- range(c(faithful[, 1], sim0[, 2], sim1[, 2]))
ylim <- range(c(faithful[, 2], sim0[, 3], sim1[, 3]))
mclust2Dplot(data = sim0[, -1], parameters = mod$parameters,
             classification = sim0[, 1], xlim = xlim, ylim = ylim)
mclust2Dplot(data = sim1[, -1], parameters = mod$parameters,
             classification = sim1[, 1], xlim = xlim, ylim = ylim)
```

Note that sim() produces a dataset in which the first column is the true classification:

```
head(sim0)
##       group    x1     x2
## [1,]      1 4.1376 71.918
## [2,]      3 4.4703 77.198
## [3,]      3 4.5090 74.671
## [4,]      2 1.9596 49.380
## [5,]      1 4.0607 78.773
## [6,]      3 4.3597 82.190
```

7.5   Large Datasets

Gaussian mixture modeling may in practice be too slow to be applied directly to datasets having a large number of observations or cases. In order to extend the method to larger datasets, both Mclust() and mclustBIC() include a provision for using a subsample of the data in the initial hierarchical clustering phase before applying the EM algorithm to the full dataset. Other functions in the **mclust** package also have this feature. Some alternatives for handling large datasets are discussed by Wehrens et al. (2004) and Fraley et al. (2005).

Starting with version 5.4 of **mclust**, a subsample of size specified by mclust.options("subset") is used in the initial hierarchical clustering phase. The default subset size is 2000. The subsampling can be removed by the command mclust.options(subset = Inf). The subset size can be specified for all the models in the current session through the subset argument to mclust.options(). Furthermore, the subsampling approach can also be implemented in the Mclust() function by setting initialization = list(subset = s), where s is a vector of logical values or numerical indices specifying the subset of data to be used in the initial hierarchical clustering phase.

EXAMPLE 7.8: Clustering on a simulated large dataset

Consider a simulated dataset of $n = 10000$ points from a three-component Gaussian mixture with common covariance matrix. First, we create a list, par, of the model parameters to be used in the simulation:

```
par <- list(
  pro = c(0.5, 0.3, 0.2),
  mean = matrix(c(0, 0, 3, 3, -4, 1), nrow = 2, ncol = 3),
  variance = sigma2decomp(matrix(c(1, 0.6, 0.6, 1.5), nrow = 2, ncol = 2),
                          G = 3))
str(par)
## List of 3
##  $ pro : num [1:3] 0.5 0.3 0.2
##  $ mean : num [1:2, 1:3] 0 0 3 3 -4 1
##  $ variance:List of 7
##  ..$ sigma : num [1:2, 1:2, 1:3] 1 0.6 0.6 1.5 1 0.6 0.6 1.5 1 0.6 ...
##  ..$ d : int 2
##  ..$ modelName : chr "EEI"
##  ..$ G : num 3
##  ..$ scale : num 1.07
##  ..$ shape : num [1:2] 1.78 0.562
##  ..$ orientation: num [1:2, 1:2] 0.555 0.832 -0.832 0.555
```

We then simulate the data with the sim() function as described in Section 7.4:

```
sim <- sim(modelName = "EEI", parameters = par, n = 10000, seed = 123)
cluster <- sim[, 1]
x <- sim[, 2:3]
```

The following code is used to fit GMMs with no subsampling, using the default subsampling, and a random sample of 1000 observations provided to the Mclust() function:

```
mclust.options(subset = Inf)  # no subsetting
system.time(mod1 <- Mclust(x))
##    user  system elapsed
## 172.981   0.936 173.866
summary(mod1$BIC)
## Best BIC values:
##                 EEI,3         EEE,3         EVI,3
## BIC       -76523.89 -76531.836803 -76541.59835
## BIC diff      0.00     -7.948006    -17.70955

mclust.options(subset = 2000)  # reset to default setting
system.time(mod2 <- Mclust(x))
##   user  system elapsed
## 14.247   0.196  14.395
summary(mod2$BIC)
## Best BIC values:
##                 EEI,3         EEE,3         EVI,3
## BIC       -76524.09 -76531.845624 -76541.89691
## BIC diff      0.00     -7.754063    -17.80535

s <- sample(1:nrow(x), size = 1000)  # one-time subsetting
system.time(mod3 <- Mclust(x, initialization = list(subset = s)))
##    user  system elapsed
## 12.091   0.146  12.197
summary(mod3$BIC)
## Best BIC values:
##               EEI,3         EEE,3         VEI,3
## BIC       -76524.17 -76532.043460 -76541.78983
## BIC diff      0.00     -7.874412    -17.62078
```

Function system.time() gives a rough indication of the computational effort associated without and with subsampling. In the latter case, using a random sample of 20% of the full set of observations, we have been able to obtain a 10 fold speedup, ending up with essentially the same final clustering model:

```
table(mod1$classification, mod2$classification)
##       1    2    3
```

```
## 1 5090    0    0
## 2    0    0 3007
## 3   10 1893    0
```

The same strategies that we have described for clustering very large datasets can also be used for classification. First, discriminant analysis with `MclustDA()` can be performed on a subset of the data. Then, the remaining data points can be classified (in reasonable sized blocks) using the associated `predict()` method.

7.6 High-Dimensional Data

Models in which the orientation is allowed to vary between components (EEV, VEV, EVV, and VVV), have $\mathcal{O}(d^2)$ parameters per cluster, where d is the dimension of the data (see Table 2.2 and Figure 2.3). For this reason, **mclust** may not work well or may otherwise be inefficient for these models when applied to high-dimensional data. It may still be possible to analyze such data with **mclust** by restricting to models with fewer parameters (spherical or diagonal models) or else by applying a dimension-reduction technique such as principal components analysis. Moreover, some of the more parsimonious models (spherical, diagonal, or fixed covariance) can be applied to datasets in which the number of observations is less than the data dimension.

7.7 Missing Data

At present, **mclust** has no direct provision for handling missing values in data. However, a function `imputeData()` for creating datasets with missing data imputations using the **mix** R package (Schafer, 2022) has been added to the **mclust** package.

The algorithm used in `imputeData()` starts with a preliminary maximum likelihood estimation step to obtain model parameters for the subset of complete observations in the data. These are used as initial values for a data augmentation step for generating posterior draws of parameters via Markov Chain Monte Carlo. Finally, missing values are imputed with simulated values drawn from the predictive distribution of the missing data given the observed data and the simulated parameters.

EXAMPLE 7.9: Imputation on the child development data

The `stlouis` dataset, available in the **mix** package, provides data from an observational study to assess the affects of parental psychological disorders on child development. Here we use imputeData() to fill in missing values in the continuous portion of the stlouis dataset; we remove the first 3 categorical variables, because **mclust** is intended for continuous variables.

```
data("stlouis", package = "mix")
x <- data.frame(stlouis[, -(1:3)], row.names = NULL)
table(complete.cases(x))
##
## FALSE  TRUE
##    42    27
apply(x, 2, function(x) prop.table(table(complete.cases(x))))
##              R1      V1      R2      V2
## FALSE 0.30435 0.43478 0.23188 0.24638
## TRUE  0.69565 0.56522 0.76812 0.75362
```

The dataset contains only 27 complete cases out of 69 observations (39%). For each variable, the percentage of missing values ranges from 23% to 43%. The pattern of missing values can be displayed graphically using the following code (see Figure 7.12):

```
library("ggplot2")
df <- data.frame(obs = rep(1:nrow(x), times = ncol(x)),
                 var = rep(colnames(x), each = nrow(x)),
                 missing = as.vector(is.na(x)))
ggplot(data = df, aes(x = var, y = obs)) +
  geom_tile(aes(fill = missing)) +
  scale_fill_manual(values = c("lightgrey", "black")) +
  labs(x = "Variables", y = "Observations") +
  theme_minimal() +
  theme(axis.text.x = element_text(margin = margin(b = 10))) +
  theme(axis.ticks.y = element_blank()) +
  theme(axis.text.y = element_blank())
```

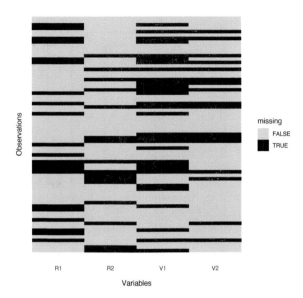

FIGURE 7.12: Image plot of the missing values pattern for the `stlouis` dataset.

A single imputation of simulated values drawn from the predictive distribution of the missing data is easily obtained as

```
ximp <- imputeData(x, seed = 123)
```

Note that the values obtained for the missing entries will vary depending on the random number seed set as argument in the `imputeData()` function call. By default, a randomly chosen seed is used.

Missing data imputations can be visualized with the function `imputePairs()` as follows:

```
imputePairs(x, ximp)
```

The corresponding pairs plot is shown in Figure 7.13.

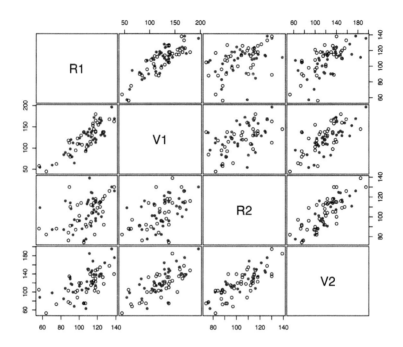

FIGURE 7.13: Pairs plot showing an instance of applying `imputeData()` to the continuous variables in the `stlouis` dataset. The open black circles correspond to non-missing data, while the filled red circles correspond to imputed missing values.

It is usually desirable to combine multiple imputations in analyses involving missing data. See Schafer (1997), Little and Rubin (2002), and van Buuren and Groothuis-Oudshoorn (2011) for details and references on multiple imputation. Besides **mix**, R packages for handling missing data include **Amelia** (Honaker et al., 2011) and **mice** (van Buuren, 2012).

EXAMPLE 7.10: Clustering with missing values on the Iris data

We now illustrate a model-based clustering analysis in the presence of missing values. Consider the `iris` data, and suppose that we want to fit a three-component model with common full covariance matrix. The model using the full set of observations is obtained as follows:

```
x <- as.matrix(iris[, 1:4])
mod <- Mclust(x, G = 3, modelNames = "EEE")
table(iris$Species, mod$classification)
##
##          1  2  3
```

```
##    setosa     50  0  0
##    versicolor  0 48  2
##    virginica   0  1 49
adjustedRandIndex(iris$Species, mod$classification)
## [1] 0.94101
mod$parameters$mean  # component means
##                 [,1]   [,2]   [,3]
## Sepal.Length 5.006 5.9426 6.5749
## Sepal.Width  3.428 2.7607 2.9811
## Petal.Length 1.462 4.2596 5.5394
## Petal.Width  0.246 1.3194 2.0254
```

Here we assume that the interest is in the final clustering partition and in the corresponding component means. Other mixture parameters, such as mixing probabilities and (co)variances, can be dealt with analogously.

Now, let us randomly designate 10% of the data values as missing:

```
isMissing <- sample(c(TRUE, FALSE), size = prod(dim(x)),
                     replace = TRUE, prob = c(0.1, 0.9))
x[isMissing] <- NA
table(cmpObs <- complete.cases(x))
##
## FALSE  TRUE
##    50   100
```

The last command shows that 100 out of 150 observations have no missing values. We can use multiple imputation to create nImp complete datasets by replacing the missing values with plausible data values. Then, an (EEE,3) clustering model is fitted to each of these datasets, and the results are pooled to get the final estimates.

A couple of issues should be mentioned. First, a well-known issue of finite mixture modeling is nonidentifiability of the components due to invariance to relabeling. This is solved in the code below using matchCluster(), which takes the first clustering result as a template for adjusting the labeling of the successive clusters. The adjusted ordering is also used for matching the other estimated parameters.

A second issue is related to how pooling of results is carried out at the end of the imputation procedure. For the clustering labels, the majority vote rule is used as implemented in function majorityVote(), whereas a simple average of estimates is calculated for pooling results for the mean parameters.

```
nImp  <- 100
muImp <- array(NA, c(ncol(x), 3, nImp))
clImp <- array(NA, c(nrow(x), nImp))
for(i in 1:nImp)
```

```
{
  xImp <- imputeData(x, verbose = FALSE)
  modImp <- Mclust(xImp, G = 3, modelNames = "EEE", verbose = FALSE)
  if (i == 1) clImp[, i] <- modImp$classification
  mcl <- matchCluster(clImp[, 1], modImp$classification)
  clImp[, i]   <- mcl$cluster
  muImp[, , i] <- modImp$parameters$mean[, mcl$ord]
}

# majority rule
cl <- apply(clImp, 1, function(x) majorityVote(x)$majority)
table(iris$Species, cl)
##             cl
##              1  2  3
##   setosa    50  0  0
##   versicolor 0 49  1
##   virginica  0  5 45
adjustedRandIndex(iris$Species, cl)
## [1] 0.88579
# pooled estimate of cluster means
apply(muImp, 1:2, mean)
##           [,1]   [,2]   [,3]
## [1,] 4.99885 5.9388 6.5667
## [2,] 3.44614 2.7684 2.9675
## [3,] 1.48357 4.2783 5.5453
## [4,] 0.25557 1.3333 2.0267
```

It is interesting to note that both the final clustering obtained by majority vote and cluster means obtained by pooling results from imputed datasets are quite close to those obtained on the original dataset.

Bibliography

Allard, D. and Fraley, C. (1997). Nonparametric maximum likelihood estimation of features in spatial point processes using Voronoï tessellation. *Journal of the American Statistical Association*, 92(440):1485–1493.

Alpaydin, E. (2014). *Introduction to Machine Learning*. MIT Press, 3rd edition.

Altman, E. I. (1968). Financial ratios, discriminant analysis and the prediction of corporate bankruptcy. *The Journal of Finance*, 23(4):589–609.

Ameijeiras-Alonso, J., Crujeiras, R. M., Rodríguez-Casal, A., The R Core Team 1996-2012, and The R Foundation 2005 (2021). *multimode: Mode Testing and Exploring*. R package version 1.5.

Anderson, E. (1935). The irises of the Gaspé Peninsula. *Bulletin of the American Iris Society*, 59:2–5.

Anderson, T. W., Olkin, I., and Underhill, L. G. (1987). Generation of random orthogonal matrices. *SIAM Journal on Scientific and Statistical Computing*, 8(4):625–629.

Arbel, J., Kon Kam King, G., Lijoi, A., Nieto-Barajas, L., and Prünster, I. (2021). BNPdensity: Bayesian nonparametric mixture modelling in R. *Australian & New Zealand Journal of Statistics*, 63(3):542–564.

Azzalini, A. and Bowman, A. W. (1990). A look at some data on the old faithful geyser. *Applied Statistics*, 39(3):357–365.

Azzalini, A. and Menardi, G. (2014). Clustering via nonparametric density estimation: The R package pdfCluster. *Journal of Statistical Software*, 57(11):1–26.

Banfield, J. and Raftery, A. E. (1993). Model-based Gaussian and non-Gaussian clustering. *Biometrics*, 49:803–821.

Barrios, E., Kon Kam King, G., Lijoi, A., Nieto-Barajas, L. E., and Prünster, I. (2021). *BNPdensity: Ferguson-Klass Type Algorithm for Posterior Normalized Random Measures*. R package version 2021.5.4.

Basford, K. E., Greenway, D. R., McLachlan, G. J., and Peel, D. (1997). Standard errors of fitted component means of normal mixtures. *Computational Statistics*, 12(1):1–18.

Bates, S., Hastie, T., and Tibshirani, R. (2021). Cross-validation: what does it estimate and how well does it do it? *arXiv preprint*. arXiv:2104.00673.

Baudry, J. P., Raftery, A. E., Celeux, G., Lo, K., and Gottardo, R. (2010). Combining mixture components for clustering. *Journal of Computational and Graphical Statistics*, 19(2):332–353.

Bensmail, H. and Celeux, G. (1996). Regularized Gaussian discriminant analysis through eigenvalue decomposition. *Journal of the American Statistical Association*, 91:1743–1748.

Biernacki, C., Celeux, G., and Govaert, G. (2000). Assessing a mixture model for clustering with the integrated completed likelihood. *IEEE Transactions on Pattern Analysis and Machine Intelligence*, 22(7):719–725.

Biernacki, C., Celeux, G., and Govaert, G. (2003). Choosing starting values for the EM algorithm for getting the highest likelihood in multivariate Gaussian mixture models. *Computational Statistics & Data Analysis*, 41(3):561–575.

Bishop, C. (2006). *Pattern Recognition and Machine Learning*. Springer-Verlag Inc, New York.

Boldea, O. and Magnus, J. R. (2009). Maximum likelihood estimation of the multivariate normal mixture model. *Journal of the American Statistical Association*, 104(488):1539–1549.

Bouveyron, C., Celeux, G., Murphy, T. B., and Raftery, A. E. (2019). *Model-Based Clustering and Classification for Data Science: With Applications in R*. Cambridge Series in Statistical and Probabilistic Mathematics. Cambridge University Press, Cambridge.

Bowman, A., to R by B. D. Ripley up to version 2.0, A. A. P., version 2.1 by Adrian Bowman, Azzalini, A., and version 2.2 by Adrian Bowman. (2022). *sm: Smoothing Methods for Nonparametric Regression and Density Estimation*. R package version 2.2-5.7.1.

Bowman, A. W. and Azzalini, A. (1997). *Applied Smoothing Techniques for Data Analysis*. Oxford: Oxford University Press.

Box, G. and Cox, D. (1964). An analysis of transformations. *Journal of the Royal Statistical Society: Series B (Statistical Methodology)*, 26(2):211–252.

Breiman, L., Friedman, J., Olshen, R., and Stone, C. J. (1984). *Classification and Regression Trees*. Wadsworth, New York.

Brier, G. W. (1950). Verification of forecasts expressed in terms of probability. *Monthly Weather Review*, 78(1):1–3.

Browne, R. P. and McNicholas, P. D. (2014). Estimating common principal components in high dimensions. *Advances in Data Analysis and Classification*, 8(2):217–226.

Byers, S. and Raftery, A. E. (1998). Nearest-neighbor clutter removal for estimating features in spatial point processes. *Journal of the American Statistical Association*, 93(442):577–584.

Campbell, J., Fraley, C., Murtagh, F., and Raftery, A. E. (1997). Linear flaw detection in woven textiles using model-based clustering. *Pattern Recognition Letters*, 18(14):1539–1548.

Campbell, J. G., Fraley, C., Stanford, D., Murtagh, F., and Raftery, A. E. (1999). Model-based methods for textile fault detection. *International Journal of Imaging Systems and Technology*, 10(4):339–346.

Cassie, R. M. (1954). Some uses of probability paper in the analysis of size frequency distributions. *Marine and Freshwater Research*, 5(3):513–522.

Celeux, G. and Govaert, G. (1995). Gaussian parsimonious clustering models. *Pattern Recognition*, 28:781–793.

Claeskens, G. and Hjort, N. L. (2008). *Model Selection and Model Averaging*. Cambridge University Press, Cambridge.

Coomans, D. and Broeckaert, I. (1986). *Potential Pattern Recognition in Chemical and Medical Decision Making*. Research Studies Press, Letchworth, England.

Coretto, P. and Hennig, C. (2016). Robust improper maximum likelihood: Tuning, computation, and a comparison with other methods for robust Gaussian clustering. *Journal of the American Statistical Association*, 111(516):1648–1659.

Csárdi, G. (2019). *cranlogs: Download Logs from the 'RStudio' 'CRAN' Mirror*. R package version 2.1.1.

Czekanowski, J. (1909). Zur Differential-Diagnose der Neadertalgruppe. *Korrespondenz-Blatt der Deutschen Gesellschaft für Anthropologie, Ethnologie, und Urgeschichte*, 40:44–47.

Dasgupta, A. and Raftery, A. E. (1998). Detecting features in spatial point processes with clutter via model-based clustering. *Journal of the American Statistical Association*, 93(441):294–302.

Davis, J. and Goadrich, M. (2006). The relationship between precision-recall and ROC curves. In *Proceedings of the 23rd International Conference on Machine Learning*, pages 233–240.

Dean, N., Murphy, T. B., and Downey, G. (2006). Using unlabelled data to update classification rules with applications in food authenticity studies. *Journal of the Royal Statistical Society: Series C (Applied Statistics)*, 55(1):1–14.

Dempster, A. P., Laird, N. M., and Rubin, D. B. (1977). Maximum likelihood from incomplete data via the EM algorithm (with discussion). *Journal of the Royal Statistical Society: Series B (Statistical Methodology)*, 39:1–38.

Dotto, F. and Farcomeni, A. (2019). Robust inference for parsimonious model-based clustering. *Journal of Statistical Computation and Simulation*, 89(3):414–442.

Dua, D. and Graff, C. (2017). UCI Machine Learning Repository.

Duong, T. (2022). *ks: Kernel Smoothing*. R package version 1.14.0.

Efron, B. (1979). Bootstrap methods: Another look at the jackknife. *Annals of Statistics*, 7:1–26.

Efron, B. (1982). *The Bootstrap, Jackknife and Other Resampling Plans*. SIAM, Philadelhia.

Escobar, M. D. and West, M. (1995). Bayesian density estimation and inference using mixtures. *Journal of the American Statistical Association*, 90(430):577–588.

Everitt, B., Landau, S., Leese, M., and Stahl, D. (2001). *Cluster Analysis*. John Wiley & Sons Ltd., Chichester, UK, 5th edition.

Ferguson, T. (1983). Bayesian density estimation by mixtures of normal distributions. In Rizvi, M. H., Rustagi, J. S., and Siegmund, D., editors, *Recent Advances in Statistics*, pages 287–302. Academic Press.

Fisher, R. A. (1936). The use of multiple measurements in taxonomic problems. *Annals of Eugenics*, 7:179–188.

Flury, B. (1988). *Common Principal Components & Related Multivariate Models*. John Wiley & Sons, Inc.

Flury, B. (1997). *A First Course in Multivariate Statistics*. Springer, New York.

Flury, B. and Riedwyl, H. (1988). *Multivariate Statistics: A Practical Approach*. Chapman & Hall Ltd.

Forina, M., Armanino, C., Castino, M., and Ubigli, M. (1986). Multivariate data analysis as a discriminating method of the origin of wines. *Vitis*, 25:189–201. Wine Recognition Database.

Forina, M., Armanino, C., Lanteri, S., and Tiscornia, E. (1983). Classification of olive oils from their fatty acid composition. In Martens, H. and Russwurm Jr., H., editors, *Food Research and Data Analysis*, pages 189–214. Applied Science Publishers, London.

Fraley, C. (1998). Algorithms for model-based Gaussian hierarchical clustering. *SIAM Journal on Scientific Computing*, 20(1):270–281.

Fraley, C. and Raftery, A. E. (1998). How many clusters? Which clustering method? Answers via model-based cluster analysis. *The Computer Journal*, 41:578–588.

Fraley, C. and Raftery, A. E. (1999). MCLUST: Software for model-based cluster analysis. *Journal of Classification*, 16(2):297–306.

Fraley, C. and Raftery, A. E. (2002). Model-based clustering, discriminant analysis, and density estimation. *Journal of the American Statistical Association*, 97(458):611–631.

Fraley, C. and Raftery, A. E. (2003). Enhanced model-based clustering, density estimation, and discriminant analysis software: MCLUST. *Journal of Classification*, 20(2):263–286.

Fraley, C. and Raftery, A. E. (2007). Bayesian regularization for normal mixture estimation and model-based clustering. *Journal of Classification*, 24(2):155–181.

Fraley, C., Raftery, A. E., Murphy, T. B., and Scrucca, L. (2012). MCLUST version 4 for R: Normal mixture modeling for model-based clustering, classification, and density estimation. Technical Report 597, Department of Statistics, University of Washington.

Fraley, C., Raftery, A. E., and Scrucca, L. (2022). *mclust: Gaussian Mixture Modelling for Model-Based Clustering, Classification, and Density Estimation*. R package version 6.0.0.

Fraley, C., Raftery, A. E., and Wehrens, R. (2005). Incremental model-based clustering for large datasets with small clusters. *Journal of Computational and Graphical Statistics*, 14(3):529–546.

Friedman, H. P. and Rubin, J. (1967). On some invariant criteria for grouping data. *Journal of the American Statistical Association*, 62(320):1159–1178.

Frühwirth-Schnatter, S. (2006). *Finite Mixture and Markov Switching Models*. Springer.

García-Escudero, L. A., Gordaliza, A., Matràn, C., and Mayo-Iscar, A. (2008). A general trimming approach to robust cluster analysis. *Annals of Statistics*, 36(3):1324–1345.

García-Escudero, L. A., Gordaliza, A., Matrán, C., and Mayo-Iscar, A. (2015). Avoiding spurious local maximizers in mixture modeling. *Statistics and Computing*, 25(3):619–633.

Giovanna, M. and Adelchi, A. (2022). *pdfCluster: Cluster Analysis via Nonparametric Density Estimation*. R package version 1.0-4.

Gnanadesikan, R. (1977). *Methods for Statistical Data Analysis of Multivariate Observations*. John Wiley & Sons, New York.

Gneiting, T. and Raftery, A. E. (2007). Strictly proper scoring rules, prediction, and estimation. *Journal of the American Statistical Association*, 102(477):359–378.

Grau, J., Grosse, I., and Keilwagen, J. (2015). PRROC: computing and visualizing precision-recall and receiver operating characteristic curves in R. *Bioinformatics*, 31(15):2595–2597.

Habbema, J. D. F., Hermans, J., and van den Broek, K. (1974). A stepwise discriminant analysis program using density estimation. In *Proceedings in Computational Statistics*, pages 101–110, Vienna: Physica-Verlag. COMPSTAT.

Habel, K., Grasman, R., Gramacy, R. B., Mozharovskyi, P., and Sterratt, D. C. (2022). *geometry: Mesh Generation and Surface Tessellation*. R package version 0.4.6.1.

Härdle, W. K. (1991). *Smoothing Techniques: With Implementation in S*. Springer Science & Business Media.

Hartigan, J. A. (1975). *Clustering Algorithms*. John Wiley & Sons, New York.

Hastie, T. and Tibshirani, R. (1996). Discriminant analysis by Gaussian mixtures. *Journal of the Royal Statistical Society: Series B (Statistical Methodology)*, 58(1):155–176.

Hastie, T., Tibshirani, R., and Friedman, J. (2009). *The Elements of Statistical Learning: Data Mining, Inference, and Prediction*. Springer-Verlag, 2nd edition.

Hathaway, R. J. (1985). A constrained formulation of maximum-likelihood estimation for normal mixture distributions. *Annals of Statistics*, 13:795–800.

Heiberger, R. M. (1978). Algorithm AS 127: Generation of random orthogonal matrices. *Journal of the Royal Statistical Society. Series C (Applied Statistics)*, 27(2):199–206.

Hennig, C. (2010). Methods for merging Gaussian mixture components. *Advances in Data Analysis and Classification*, 4(1):3–34.

Hennig, C. (2020). *fpc: Flexible Procedures for Clustering*. R package version 2.2-9.

Hennig, C. and Hausdorf, B. (2020). *prabclus: Functions for Clustering and Testing of Presence-Absence, Abundance and Multilocus Genetic Data*. R package version 2.3-2.

Honaker, J., King, G., and Blackwell, M. (2011). Amelia II: A program for missing data. *Journal of Statistical Software*, 45(7):1–47.

Hubert, L. and Arabie, P. (1985). Comparing partitions. *Journal of Classification*, 2:193–218.

Hurley, C. (2019). *gclus: Clustering Graphics*. R package version 1.3.2.

Hyndman, R. J. (1996). Computing and graphing highest density regions. *The American Statistician*, 50(2):120–126.

Ihaka, R., Murrell, P., Hornik, K., Fisher, J. C., Stauffer, R., Wilke, C. O., McWhite, C. D., and Zeileis, A. (2022). *colorspace: A Toolbox for Manipulating and Assessing Colors and Palettes*. R package version 2.0-3.

Ingrassia, S. and Rocci, R. (2007). Constrained monotone EM algorithms for finite mixture of multivariate Gaussians. *Computational Statistics & Data Analysis*, 51(11):5339–5351.

Izenman, A. J. and Sommer, C. J. (1988). Philatelic mixtures and multimodal densities. *Journal of the American Statistical Association*, 83(404):941–953.

Kass, R. E. and Raftery, A. E. (1995). Bayes factors. *Journal of the American Statistical Association*, 90:773–795.

Keilwagen, J., Grosse, I., and Grau, J. (2014). Area under precision-recall curves for weighted and unweighted data. *PLOS ONE*, 9(3).

Keribin, C. (2000). Consistent estimation of the order of mixture models. *Sankhya Series A*, 62(1):49–66.

Kohavi, R. (1995). A study of cross-validation and bootstrap for accuracy estimation and model selection. In *Proceedings of the 14th International Joint Conference on Artificial Intelligence - Volume 2*, IJCAI'95, pages 1137–1143, San Francisco, CA, USA. Morgan Kaufmann Publishers Inc.

Konishi, S. and Kitagawa, G. (2008). *Information Criteria and Statistical Modeling*. Springer Science & Business Media.

Kruppa, J., Liu, Y., Biau, G., Kohler, M., König, I. R., Malley, J. D., and Ziegler, A. (2014a). Probability estimation with machine learning methods for dichotomous and multicategory outcome: Theory. *Biometrical Journal*, 56(4):534–563.

Kruppa, J., Liu, Y., Diener, H.-C., Holste, T., Weimar, C., König, I. R., and Ziegler, A. (2014b). Probability estimation with machine learning methods for dichotomous and multicategory outcome: Applications. *Biometrical Journal*, 56(4):564–583.

Kuhn, M. and Johnson, K. (2013). *Applied Predictive Modeling*. Springer, New York.

Langrognet, F., Lebret, R., Poli, C., Iovleff, S., Auder, B., and Iovleff, S. (2022). *Rmixmod: Classification with Mixture Modelling*. R package version 2.1.7.

Lantz, B. (2019). *Machine Learning with R: Expert Techniques for Predictive Modeling*. Packt Publishing, 3rd edition.

Lazarsfeld, P. F. (1950a). The logical and mathematical foundation of latent structure analysis. In Stouffer, S. A., editor, *Measurement and Prediction, Volume IV of The American Soldier: Studies in Social Psychology in World War II*, chapter 10. Princeton University Press.

Lazarsfeld, P. F. (1950b). The logical and mathematical foundation of latent structure analysis. In Stouffer, S. A., editor, *Measurement and Prediction*, pages 362–412. Princeton University Press.

Little, R. J. and Rubin, D. B. (2002). *Statistical Analysis with Missing Data*. John Wiley & Sons, 2nd edition.

Loader, C. (1999). *Local Regression and Likelihood*. Springer Verlag, New York.

Maechler, M., Rousseeuw, P., Struyf, A., Hubert, M., and Hornik, K. (2022). *cluster: Cluster Analysis Basics and Extensions*. R package version 2.1.4.

Mangasarian, O. L., Street, W. N., and Wolberg, W. H. (1995). Breast cancer diagnosis and prognosis via linear programming. *Operations Research*, 43(4):570–577.

Marron, J. S. and Wand, M. P. (1992). Exact mean integrated squared error. *Annals of Statistics*, 20(2):712–736.

McLachlan, G. (2004). *Discriminant Analysis and Statistical Pattern Recognition*. John Wiley & Sons, New York.

McLachlan, G. and Krishnan, T. (2008). *The EM Algorithm and Extensions*. Wiley-Interscience, Hoboken, New Jersey, 2nd edition.

McLachlan, G. J. (1977). Estimating the linear discriminant function from initial samples containing a small number of unclassified observations. *Journal of the American Statistical Association*, 72(358):403–406.

McLachlan, G. J. (1987). On bootstrapping the likelihood ratio test statistic for the number of components in a normal mixture. *Applied Statistics*, 36:318–324.

McLachlan, G. J. and Basford, K. E. (1988). *Mixture Models: Inference and Applications to Clustering*. Marcel Dekker Inc., New York.

McLachlan, G. J. and Peel, D. (2000). *Finite Mixture Models*. Wiley, New York.

McLachlan, G. J. and Rathnayake, S. (2014). On the number of components in a Gaussian mixture model. *Wiley Interdisciplinary Reviews: Data Mining and Knowledge Discovery*, 4(5):341–355.

McNeil, D. R. (1977). *Interactive Data Analysis*. Wiley, New York.

McNicholas, P. D. (2016). *Mixture Model-Based Classification*. CRC Press.

McNicholas, P. D., ElSherbiny, A., McDaid, A. F., and Murphy, T. B. (2022). *pgmm: Parsimonious Gaussian Mixture Models*. R package version 1.2.6.

Meng, X.-L. and Rubin, D. B. (1991). Using EM to obtain asymptotic variance-covariance matrices: The SEM algorithm. *Journal of the American Statistical Association*, 86(416):899–909.

Morris, K., McNicholas, P., and Scrucca, L. (2013). Dimension reduction for model-based clustering via mixtures of multivariate t-distributions. *Advances in Data Analysis and Classification*, 7(3):321–338.

Morris, K. and McNicholas, P. D. (2013). Dimension reduction for model-based clustering via mixtures of shifted asymmetric Laplace distributions. *Statistics & Probability Letters*, 83(9):2088–2093.

Morris, K. and McNicholas, P. D. (2016). Clustering, classification, discriminant analysis, and dimension reduction via generalized hyperbolic mixtures. *Computational Statistics & Data Analysis*, 97:133–150.

Murtagh, F. and Raftery, A. E. (1984). Fitting straight lines to point patterns. *Pattern Recognition*, 17(5):479–483.

Neath, A. A. and Cavanaugh, J. E. (2012). The Bayesian information criterion: background, derivation, and applications. *Wiley Interdisciplinary Reviews: Computational Statistics*, 4(2):199–203.

Newton, M. A. and Raftery, A. E. (1994). Approximate Bayesian inference with the weighted likelihood bootstrap (with discussion). *Journal of the Royal Statistical Society: Series B (Statistical Methodology)*, 56:3–48.

Ogle, D. (2022). *FSAdata: Fisheries Stock Analysis, Datasets*. R package version 0.4.0.

O'Hagan, A., Murphy, T. B., Scrucca, L., and Gormley, I. C. (2019). Investigation of parameter uncertainty in clustering using a Gaussian mixture model via jackknife, bootstrap and weighted likelihood bootst rap. *Computational Statistics*, 34(4):1779–1813.

Okabe, M. and Ito, K. (2008). Color universal design (CUD) - how to make figures and presentations that are friendly to colorblind people. *J* Fly: Data Depository for Drosophila Researchers*.

O'Neill, T. J. (1978). Normal discrimination with unclassified observations. *Journal of the American Statistical Association*, 73(364):821–826.

Peel, D. and McLachlan, G. J. (2000). Robust mixture modelling using the *t* distribution. *Statistics and Computing*, 10(4):339–348.

Posse, C. (2001). Hierarchical model-based clustering for large datasets. *Journal of Computational and Graphical Statistics*, 10(3):464–486.

Punzo, A. and McNicholas, P. D. (2016). Parsimonious mixtures of multivariate contaminated normal distributions. *Biometrical Journal*, 58(6):1506–1537.

R Core Team (2022). *R: A Language and Environment for Statistical Computing*. R Foundation for Statistical Computing, Vienna, Austria.

Raftery, A. E. and Dean, N. (2006). Variable selection for model-based clustering. *Journal of the American Statistical Association*, 101(473):168–178.

Reaven, G. and Miller, R. (1979). An attempt to define the nature of chemical diabetes using a multidimensional analysis. *Diabetologia*, 16(1):17–24.

Richardson, S. and Green, P. J. (1997). On Bayesian analysis of mixtures with an unknown number of components (with discussion). *Journal of the Royal Statistical Society: Series B (Statistical Methodology)*, 59(4):731–792.

Ripley, B. (2022). *MASS: Support Functions and Datasets for Venables and Ripley's MASS*. R package version 7.3-58.1.

Roeder, K. (1990). Density estimation with confidence sets exemplified by superclusters and voids in the galaxies. *Journal of the American Statistical Association*, 85(411):617–624.

Roeder, K. and Wasserman, L. (1997). Practical Bayesian density estimation using mixtures of normals. *Journal of the American Statistical Association*, 92(439):894–902.

Saerens, M., Latinne, P., and Decaestecker, C. (2002). Adjusting the outputs of a classifier to new a priori probabilities: a simple procedure. *Neural Computation*, 14(1):21–41.

Schafer, J. L. (1997). *Analysis of Incomplete Multivariate Data*. Chapman & Hall/CRC, London.

Schafer, J. L. (2022). *mix: Estimation/Multiple Imputation for Mixed Categorical and Continuous Data*. R package version 1.0-11.

Schwartz, G. (1978). Estimating the dimension of a model. *Annals of Statistics*, 6:31–38.

Scott, A. and Symons, M. J. (1971). Clustering methods based on likelihood ratio criteria. *Biometrics*, 27(2):387–397.

Scott, D. W. (2009). *Multivariate Density Estimation: Theory, Practice, and Visualization*. John Wiley & Sons, 2nd edition.

Scrucca, L. (2010). Dimension reduction for model-based clustering. *Statistics and Computing*, 20(4):471–484.

Scrucca, L. (2014). Graphical tools for model-based mixture discriminant analysis. *Advances in Data Analysis and Classification*, 8(2):147–165.

Scrucca, L. (2016). Identifying connected components in Gaussian finite mixture models for clustering. *Computational Statistics & Data Analysis*, 93:5–17.

Scrucca, L. (2019). A transformation-based approach to Gaussian mixture density estimation for bounded data. *Biometrical Journal*, 61(4):1–16.

Scrucca, L. (2022). *mclustAddons: Addons for the 'mclust' Package*. R package version 0.7.2.

Scrucca, L., Fop, M., Murphy, T. B., and Raftery, A. E. (2016). mclust 5: clustering, classification and density estimation using Gaussian finite mixture models. *The R Journal*, 8(1):205–233.

Scrucca, L. and Raftery, A. E. (2015). Improved initialisation of model-based clustering using Gaussian hierarchical partitions. *Advances in Data Analysis and Classification*, 4(9):447–460.

Scrucca, L. and Serafini, A. (2019). Projection pursuit based on Gaussian mixtures and evolutionary algorithms. *Journal of Computational and Graphical Statistics*, 28(4):847–860.

Silverman, B. W. (1998). *Density Estimation for Statistics and Data Analysis*. Chapman & Hall/CRC.

Simonoff, J. S. (1996). *Smoothing Methods in Statistics*. Springer.

Sing, T., Sander, O., Beerenwinkel, N., and Lengauer, T. (2005). ROCR: visualizing classifier performance in R. *Bioinformatics*, 21(20):7881.

Stahl, D. and Sallis, H. (2012). Model-based cluster analysis. *Wiley Interdisciplinary Reviews: Computational Statistics*, 4(4):341–358.

Street, W. N., Wolberg, W. H., and Mangasarian, O. L. (1993). Nuclear feature extraction for breast tumor diagnosis. In *Biomedical Image Processing and Biomedical Visualization*, volume 1905, pages 861–870. International Society for Optics and Photonics.

Titterington, D. M., Smith, A. F., and Makov, U. E. (1985). *Statistical Analysis of Finite Mixture Distributions*. John Wiley & Sons, Chichester; New York.

Todorov, V. (2022). *rrcov: Scalable Robust Estimators with High Breakdown Point*. R package version 1.7-2.

Tortora, C., ElSherbiny, A., Browne, R. P., Franczak, B. C., , McNicholas, P. D., and Amos., D. D. (2022). *MixGHD: Model Based Clustering, Classification and Discriminant Analysis Using the Mixture of Generalized Hyperbolic Distributions*. R package version 2.3.7.

Unwin, A. (2015). *GDAdata: Datasets for the Book Graphical Data Analysis with R*. R package version 0.93.

van Buuren, S. (2012). *Flexible Imputation of Missing Data*. Chapman & Hall/CRC.

van Buuren, S. and Groothuis-Oudshoorn, K. (2011). mice: Multivariate imputation by chained equations in r. *Journal of Statistical Software*, 45(3):1–67.

Van Engelen, J. E. and Hoos, H. H. (2020). A survey on semi-supervised learning. *Machine Learning*, 109(2):373–440.

Venables, W. N. and Ripley, B. D. (2013). *Modern Applied Statistics with S-PLUS*. Springer Science and Business Media.

Wand, M. (2021). *KernSmooth: Functions for Kernel Smoothing Supporting Wand & Jones (1995)*. R package version 2.23-20.

Wang, N., Raftery, A., and Fraley, C. (2017). *covRobust: Robust Covariance Estimation via Nearest Neighbor Cleaning*. R package version 1.1-3.

Wang, N. and Raftery, A. E. (2002). Nearest neighbor variance estimation (NNVE): Robust covariance estimation via nearest neighbor cleaning (with discussion). *Journal of the American Statistical Association*, 97(460):994–1019.

Ward, J. H. (1963). Hierarchical grouping to optimize an objective function. *Journal of the American Statistical Association*, 58(301):236–244.

Wehrens, R., Buydens, L., Fraley, C., and Raftery, A. E. (2004). Model-based clustering for image segmentation and large datasets via sampling. *Journal of Classification*, 21(2):231–253.

Wickham, H. (2016). *ggplot2: Elegant Graphics for Data Analysis*. Springer-Verlag, New York, 2nd edition.

Wickham, H., Averick, M., Bryan, J., Chang, W., McGowan, L. D., Francois, R., Grolemund, G., Hayes, A., Henry, L., Hester, J., Kuhn, M., Pedersen, T. L., Miller, E., Bache, S. M., Müller, K., Ooms, J., Robinson, D., Seidel, D. P., Spinu, V., Takahashi, K., Vaughan, D., Wilke, C., Woo, K., and Yutani, H. (2019). Welcome to the tidyverse. *Journal of Open Source Software*, 4(43):1686.

Wickham, H. and Cook, D. (2022). *tourr: Implement Tour Methods in R Code*. R package version 0.6.2.

Wickham, H. and Henry, L. (2022). *tidyr: Tidy Messy Data*. R package version 1.2.1.

Wilkinson, L. (2005). *The Grammar of Graphics*. Springer-Verlag, New York.

Wolfe, J. H. (1963). *Object Cluster Analysis of Social Areas*. PhD thesis, University of California, Berkeley.

Wolfe, J. H. (1965). A computer program for the maximum likelihood analysis of types. USNPRA Technical Bulletin 65-15, U.S. Naval Personnel Research Activity, San Diego, CA.

Wolfe, J. H. (1967). NORMIX: Computational methods for estimating the parameters of multivariate normal mixtures of distributions. Technical report, Naval Personnel Research Activity San Diego CA.

Wolfe, J. H. (1970). Pattern clustering by multivariate mixture analysis. *Multivariate Behavioral Research*, 5:329–350.

Wong, B. (2011). Points of view: Color blindness. *Nature Methods*, 8(441).

Wu, C. J. (1983). On the convergence properties of the EM algorithm. *Annals of Statistics*, 11(1):95–103.

Zeileis, A., Fisher, J. C., Hornik, K., Ihaka, R., McWhite, C. D., Murrell, P., Stauffer, R., and Wilke, C. O. (2020). colorspace: A toolbox for manipulating and assessing colors and palettes. *Journal of Statistical Software*, 96(1):1–49.

Zhu, X. and Goldberg, A. B. (2009). *Introduction to Semi-Supervised Learning*, volume 3 of *Synthesis Lectures on Artificial Intelligence and Machine Learning*. Morgan & Claypool Publishers.

Index

For Product Safety Concerns and Information please contact our
EU representative GPSR@taylorandfrancis.com Taylor & Francis
Verlag GmbH, Kaufingerstraße 24, 80331 München, Germany